MULTIPLE REGRESSION

To our Teachers

B. J. Winer

Quinn McNemar

MULTIPLE REGRESSION: Testing and Interpreting Interactions

Leona S. Aiken
Stephen G. West
Arizona State University

With contributions by Raymond R. Reno
University of Notre Dame

For information address:

SAGE Publications, Inc.
2455 Teller Road
Thousand Oaks, California 91320
E-mail: order@sagepub.com

SAGE Publications Ltd.
6 Bonhill Street
London EC2A 4PU
United Kingdom

SAGE Publications India Pvt. Ltd.
M-32 Market
Greater Kailash I
New Delhi 110 048 India

Printed in the United States of America

Library of Congress Cataloging-in-Publication Data

Aiken, Leona S.
Multiple regression: Testing and interpreting interactions / Leona S. Aiken and Stephen G. West.
p. cm.
Includes bibliographical references and index.
ISBN 0-8039-3605-2 (c) ISBN 0-7619-0712-2 (pbk.)
1. Regression analysis. I. West, Stephen G. II. Title.
QA278.2.A34 1991
519.5′36—dc20 91-2062

This book is printed on acid-free paper.

96 97 98 99 01 02 03 10 9 8 7 6 5 4

First paperback printing 1996

Sage Production Editor: Susan McElroy

Contents

Preface

Social scientists have long had interests in studying interactions between variables. Whether hypothesized directly by theory or resulting from an attempt to establish a boundary condition for a relationship, the testing of interactions has formed an important part of their research.

In 1984 we began working on a number of research projects that involved complex interactions between continuous variables. Although several good articles and small sections of textbooks addressed interactions in multiple regression models, none provided what could be considered to be a comprehensive treatment to which we could refer students and colleagues. This situation contrasted sharply to the specification of procedures for examining interactions between categorical variables. Procedures for testing, graphically displaying, and post hoc probing of interactions between categorical variables for virtually all possible realistic designs have for years been laboriously detailed in Analysis of Variance textbooks. The prevalence of incomplete or nonoptimal treatments of continuous variable interactions in several major journals in psychology further confirmed our observation that psychologists did not have clearcut guidelines for the analysis of interactions between continuous variables. Indeed, the usual practice in 1984 was either to ignore continuous variable interaction effects or to follow the nonoptimal procedure of doing a median split of the continuous variables followed by an Analysis of Variance. Thus the need for a comprehensive source on the treatment of interactions in multiple regression seemed clear. Such a sourcebook would be useful to both graduate students and researchers facing this statistical problem.

This book provides clear prescriptions for the probing and interpretation of continuous variable interactions that are the analogs of existing prescriptions for categorical variable interactions. We provide prescriptions for probing and interpreting two- and three-way continuous variable interactions, including those involving nonlinear components. The interaction of continuous and categorical variables, the hallmark of analysis of covariance and related procedures, is treated as a special case of our general prescriptions. The issue of power of tests for continuous variable interactions, and the impact of measurement error on power are also addressed. Simple approaches for operationalizing the prescriptions for post hoc tests of interactions with standard statistical computer packages are provided.

The text is designed for researchers and graduate students who are familiar with multiple regression analysis involving simple linear relationships of a set of continuous predictors to a criterion. Hence the material is accessible to typical lower level graduate students in the social sciences, education, and business. The text can usefully serve as a supplement to introductory graduate level statistics courses. The required mathematical background is high school algebra. Although there are smatterings of calculus and matrix algebra, readers unfamiliar with these mathematical forms will not be disadvantaged in their understanding of the material or the application of the prescriptions to their own research.

Several individuals have made significant contributions to the development of this monograph. Most notable are the major contributions of Ray Reno. He provided all the simulations and computer analyses we have reported throughout the text. His computer examples render our prescriptions accessible to anyone with knowledge of the regression analysis programs in standard statistical computer packages. We also appreciate the input of a number of individuals who read and commented on versions of the manuscript, among them Sanford Braver, Patrick Curran, Joseph Hepworth, and Jenn-Yun Tein of Arizona State University; Richard Darlington of Cornell University, Charles Judd of the University of Colorado, and Herbert Marsh of the University of Western Sydney, Macarthur. David Kenny of the University of Connecticut and James Jaccard of the State University of New York, Albany provided thorough reviews of the manuscript; their helpful input is strongly reflected in the final form of the text. Special acknowledgment is due to David Kenny and Patrick Curran, who provided extremely detailed and probing commentaries. We also thank Susan Maxwell for her many insights about interactions. Finally, we are very appreciative of the encouragement, guidance, flexibility, and patience of our Sage Editor, Deborah Laughton, and the painstaking efforts of production editor Susan McElroy in preparing the book for publication.

The clerical and editorial efforts of Jane Hawthorne and Kathy Sidlik during various stages of the project are gratefully acknowledged. Andrea Fenaughty provided greatly appreciated assistance with the indexing and referencing. Support for graduate assistant Ray Reno, as well as Jane Hawthorne and Kathy Sidlik, was provided by the College of Liberal Arts and Sciences, Arizona State University. The efforts of Steve West were in part supported by National Institute of Mental Health grant P50MH39246.

L. S. A. and S. G. W.
Tempe, January 1991

1 Introduction

This book is concerned with a statistical problem commonly faced by researchers in the social sciences, business, education, and communication: How to structure, test, and interpret complex multiple regression models containing interactions, curvilinear effects, or both. To understand this problem, consider the following example. A researcher wishes to investigate the effects of life stress (X) and the amount of social support received by the person (Z) on physical illness (Y). He obtains highly reliable continuous measures of each of the variables on a large sample of subjects. Following the dictates of what has become "standard practice" in many research areas, he enters the two predictor variables and the outcome variable into any of the standard regression packages and estimates the familiar regression equation reproduced below:

$$\hat{Y} = b_1X + b_2Z + b_0 \tag{1.1}$$

The tests of the b_1 and b_2 coefficients are easily accomplished and inform the researcher whether stress and social support, respectively, have a nonzero linear relationship to physical illness in the population. The b_0 coefficient, which will appear as the final term in all regression equations in this book, represents the regression constant or intercept and will only rarely be of theoretical interest. $\hat{Y}$ represents the predicted value of Y.

But, is this the regression model that the researcher really wished to test? Like many other social scientists, the researcher's interest was *not* so much in whether stress and social support have the linear and additive effects on physical illness specified by equation 1.1. Rather, in line with prior theorizing (Cobb, 1976; LaRocco, House, & French, 1980), he wished to test the explicit hypothesis that social support buffers the effect of stress on physical illness. That is, the researcher predicted that while each individual's level of stress is positively related to his or her level of physical illness, the strength of this relationship weakens as the level of social support received by the individual increases. Indeed, for individuals with very high levels of social support, the researcher would predict that there would be little, if any, relationship between stress and physical health. This hypothesis implies that there should be an interaction of stress and social support in predicting physical illness, a relationship represented by equation 1.2 below:

$$\hat{Y} = b_1X + b_2Z + b_3XZ + b_0 \tag{1.2}$$

Thus equation 1.1 does *not* test the researcher's hypothesis.

Substantive theory such as the stress-buffering model described above often specifies that the value of an outcome variable depends jointly upon the value of two or more predictor variables. Also, in empirical work in a new research area investigators often initially attempt to find general causal relationships of the form X causes Y. When they have established such relationships, they then attempt to specify conditions under which the causal relationship is weakened (moderated) or strengthened (amplified). These are interactions.

Interest in complex hypotheses that are not adequately represented by simple, additive linear regression equations is common in many disciplines. These complex hypotheses include not only interactions but also curvilinear relationships between the predictor variables and the outcome. The following examples selected from a variety of areas in the social sciences, business, and education are illustrative. In each case, the regression equations must be structured to contain higher order terms representing interactions, curvilinear effects, or both in order to test the researcher's hypothesis properly.

1. The number of years of job experience is positively related to the worker's salary. However, this relationship may be moderated by the percentage of female workers in the occupation: Occupations with a higher percentage of female workers are predicted to have only small increases in salary as a function of job experience relative to occupations with a lower percentage of female workers (England, Farkas, Kilbourne, & Dou, 1988).

2. A number of researchers have hypothesized that superior classroom performance results when the student's personal style matches the nature of the classroom environment (see Cronbach & Snow, 1977). As one illustration, Domino (1968, 1971) hypothesized that the student's degree of personal independence (versus conformity) would interact with the directiveness of the instructor (encourages independence versus encourages conformity) to predict the student's grade in the course.

3. Studies of workspace design find that the larger the number of people in each office, the higher the rate of turnover of personnel in the office. However, this relationship is expected to be weakened to the extent each employee's desk is enclosed by partitions (Oldham & Fried, 1987).

4. Several theories of leadership propose that the group's performance is a complex function of both the leader's style and the nature of the situation. For example, Fiedler (1967; Fiedler, Chemers, & Mahar, 1976) proposed that in favorable situations (defined as having high task structure, good leader–member relations, high leader power) and in very unfavorable situations (low values on each of these situational characteristics), leaders with a task-oriented style would elicit the highest level of performance from their groups. However, in mixed (moderate) situations, leaders with a task-oriented style would elicit the *lowest* level of performance.

5. In a study of vetos of legislation by U.S. presidents, Simonton (1987) hypothesized that the degree of success in sustaining the veto would reflect an interaction between the president's personal level of flexibility and the percentage of the electoral college vote he received in the previous election.

6. A classic "law" in psychology (Yerkes & Dodson, 1908) hypothesizes that performance will show an inverted-U shaped relationship to the person's level of physiological arousal. The point of maximum performance and the exact shape of the curve are determined by the difficulty of the task.

The general methods of structuring complex regression equations to test such hypotheses explicitly were first proposed over two decades ago in the social sciences. Cohen (1968) proposed multiple regression (MR) analysis as a general data analytic strategy. According to this strategy, any combination of categorical and continuous variables can be analyzed within a multiple regression (MR) framework simply through the appropriate dummy coding of the categorical variables. Interactions can be represented as product terms, and curvilinear relationships can be represented through higher order terms in the regression equation. Other early proposals for the structuring and testing of complex regression models involving interactions and/or higher order effects were made in sociology by Allison (1977), Blalock (1965), and Southwood (1978) and in political science by Friedrich (1982) and Wright (1976).

Despite the availability of general procedures for testing interactions and curvilinear effects within an MR framework, the actual practice of researchers in many areas of social science, business, and education indicates these strategies have only rarely been followed. Many researchers, as in our example of stress, social support, and physical illness above, have incorrectly utilized simple additive regression models that ignore possible interaction effects.

Other researchers, originally trained in the use of analysis of variance (ANOVA) models, have frequently utilized median splits of their continuous variables. This practice does allow data to be subjected to the familiar procedures of the 2 × 2 factorial ANOVA. Unfortunately, this practice is also associated with substantial costs. Median splits of continuous variables throw away information, reducing the power of the statistical test: They make it much more difficult to detect significant effects when in fact they do exist (Cohen, 1983). For an interaction effect of any specified magnitude, a substantially larger sample of subjects will be needed when the median split approach is used rather than the MR with interactions approach in order to achieve adequate statistical power. The median split approach may often be less informative in practice than MR approaches if higher order relationships between the predictors and criterion exist. Finally, the MR approach uses all of the information available in the predictor variables to provide direct estimates of the effect size and percentage of variance accounted for.[1]

As an illustration of the magnitude of the problem of the nonutilization of MR with interactions in one social science, we conducted a survey of four psychological journals that frequently publish articles involving analyses of multiple continuous predictor (independent) variables.[2] Three categories of analysis strategies for continuous predictors were tabulated: (a) ANOVA with continuous variables broken into categories (ANOVA with cutpoints, nearly always median splits), (b) MR without interactions, and (c) MR with interactions. A total of the 148 articles involving the analysis of two (or more) continuous variables were located: In 77% of these articles one of the first two strategies was used rather than MR with interactions. Although we have not undertaken a formal survey of journals in other social science disciplines, education, or business, our impression is that the use of complex MR models with interactions continues to be rare relative to analysis strategies (a) and (b).

Over two decades have passed since the initial proposals of MR as a general data analytic strategy in the social sciences by Blalock (1965) and Cohen (1968). Other works have also periodically echoed Cohen's message (e.g., Cohen & Cohen, 1975, 1983; Darlington, 1990; Kenny, 1985; Neter, Wasserman, & Kutner, 1989; Pedhazur, 1982). Why have researchers been so slow to utilize these techniques in the analysis of studies involving two or more predictor variables? We believe this underutilization of MR approaches stems in large part from several impediments that arise when researchers actually attempt to utilize the general procedures that have been outlined and to interpret their results. The purpose of this book is to provide a detailed explanation of the procedures through which regression equations containing interactions and higher order nonlinear terms may be structured, tested, and interpreted. In pursuing this aim, we present an integration of recent work in the psychological, sociological, and statistical literatures that clarifies a number of earlier points of confusion in the literature that have also served as impediments to the use of the MR approach.

Chapter 2 addresses a number of issues involved in the interpretation of interactions between two *continuous* variables in MR. One major impediment to the use of MR has been that procedures for displaying and probing significant interactions have not been readily available. That is, once an interaction was found to be significant, exactly what should one do next? In this chapter, we present the graphical approaches to examining the interaction in equation 1.2 originally developed by Cohen and Cohen (1975, 1983), present analyses to answer questions about the form (ordinal versus disordinal) of the interaction, and derive procedures for post hoc statistical probing of interactions between continuous variables that closely parallel simple effects testing within the ANOVA framework.

Chapter 3 addresses another impediment to the use of MR with interaction terms: the lack of invariance of the MR results even under simple linear transformations of the data. To understand this problem, consider analyzing a data set using equation 1.2 that contains first order terms for X and Z and the linear interaction of X and Z:

$$\hat{Y} = b_1X + b_2Z + b_3XZ + b_0$$

Two analyses are conducted: First, the data are analyzed in raw score form; second, the data are analyzed with X and Z *centered* (put in deviation score form), with the interaction created from the deviation score forms of X and Z. The results of these two analyses would differ, perhaps dramatically. Only the b_3 coefficient for the interaction term would remain the same in both equations (see Cohen, 1978). Such shifts in the results of the analyses of regression equations containing interactions or higher order variables under transformation are disturbing. This problem has led to substantial discussion in the social science literature (see, e.g., Friedrich, 1982, in political science; Cohen, 1978, and Sockloff, 1976, in psychology; Althauser, 1971, and Arnold & Evans, 1979, in sociology). Confusion and conflicting recommendations have resulted (e.g., Schmidt, 1973, versus Sockloff, 1976). Chapter 3 clarifies the source of the problem of failure to maintain invariance. Procedures through which researchers may work with equations containing higher order terms and maintain unambiguous interpretations of their effects are highlighted. The interpretation of all regression coefficients in equations containing interactions is explained. Finally, a standardized solution for equations containing interactions is presented.

Chapter 4 and 5 examine problems in testing interactions in more complex regression equations. Most discussions of interactions in MR have focused exclusively on the simple model involving two predictors and their interaction represented by equation 1.2. Chapter 4 generalizes the procedures for the treatment of interactions to the three predictor case. Methods for graphically displaying the three-way interaction and for conducting post hoc tests that are useful in probing the nature of the interaction are discussed.

Chapter 5 considers several complications that arise in structuring, testing, and interpreting regression equations containing higher order terms to represent curvilinear (quadratic) effects and their complex interactions. The methods for graphically portraying and conducting post hoc tests of interactions developed in earlier chapters are generalized to a variety of complex regression equations.

Chapters 5 and 6 also address another impediment to the use of complex MR models for researchers trained in the ANOVA tradition. In equation 1.2 where the XZ term with one degree of freedom (df) represents fully the interaction between X and Z, generalizing from ANOVA to regression is relatively easy. This generalization is less straightforward when interactions have more than one degree of freedom. In ANOVA there is always one source of variation for the whole interaction and one omnibus test for significance even when an interaction has several degrees of freedom. However, in more complex MR equations, a series of terms, each with one degree of freedom, may represent the interaction. For example, terms representing the linear X by linear Z component (XZ) and the curvilinear (quadratic) X by linear Z component (X^2Z) of the interaction between *X* and *Z* may need to be built into the regression equation. In such cases, the generalization from one source of variation and one omnibus test for the whole interaction in ANOVA to the MR framework is not generally familiar to researchers. Similar problems exist for testing multiple df "main effects" in MR in the absence of an interaction. Chapter 5 provides guidelines for the structuring and interpreting of these more complex regression equations.

Chapter 6 further extends the consideration of this issue by developing a variety of procedures for model and effect testing in complex regression equations. Strategies for testing and interpreting lower order effects in MR are developed. Global tests of a variety of hypotheses (e.g., the overall linearity of regression) that are based on sets of terms in the equation are discussed. Hierarchical effect-by-effect and term-by-term step-down testing strategies are presented for simplifying complex regression equations. Procedures are presented for identifying the scale-independent term(s) that can be legitimately tested at each step, yielding proper reduced equations that can be interpreted.

Chapter 7 generalizes our treatment of interactions to cases involving combinations of categorical and continuous predictor variables. Issues arising in the representation of categorical variables and the interpretation of the regression coefficients are discussed. Post hoc tests of the interaction are presented that examine differences between regression equations for the groups defined by the categorical variable.

Chapter 8 addresses the problem of measurement error in the predictor variables and its effect on interactions. Several methods of correcting for measurement error are presented, and their performance evaluated. The dramatic effect of measurement error on statistical power (the ability of statistical tests to detect true interactions) is shown.

Finally, Chapter 9 briefly contrasts the ANOVA and MR approaches as they have been used in practice. ANOVA was classically applied to experimental designs, whereas multiple regression was applied to measured variables. We explore some of the areas in ANOVA and MR in which lingering traditions have led to divergent practices: model specification, functional form, and examination of the tenability of assumptions.

We hope that this book will introduce investigators in a variety of social science disciplines to the major issues involved in the design and analysis of research involving one or more continuous predictor variables. We also hope the book will provide an increased understanding of interactions in multiple regression and will help remove the impediments to the use of MR as a general data analytic strategy.

Notes

1. The ANOVA model is appropriate when (a) the levels of the predictor variable are discrete rather than continuous or (b) the relationship between the predictor and outcome variables is a step function rather than linear or curvilinear (Kenny, 1985). As Cohen and Cohen (1983; see also Chapter 7 of the present book) have shown, these cases can also be equally well represented using the MR approach.

2. When beginning this project, we reviewed the 1984 volumes of four major journals of the American Psychological Association: *Journal of Abnormal Psychology, Developmental Psychology, Journal of Consulting and Clinical Psychology*, and *Journal of Personality and Social Psychology*. These journals were selected because they are leading journals in psychology and they most frequently publish articles involving analyses of multiple continuous predictor (independent) variables. Our estimate that 23% of the articles involving two (or more) continuous variables used MR with interactions may be considered to reflect the quality of the best practice rather than average level of practice in these areas of psychology in 1984.

2 Interactions Between Continuous Predictors in Multiple Regression

In this chapter we begin by explaining what the interaction between two continuous predictors in a regression analysis signifies about the relationship of the predictors to the criterion. We then address the problem of displaying and statistically probing interactions between continuous variables in MR for the simple case in which there is only one term (b_3XZ) involving both X and Z. This case is represented by equation 2.1:

$$\hat{Y} = b_1X + b_2Z + b_3XZ + b_0 \quad (2.1)$$

For ease of presentation the exposition in this chapter assumes that the single predictor variables (here X and Z) have been *centered* (i.e., put in deviation score form so that their means are zero) and that the XZ term has been formed by multiplying together the two centered predictors. Centering variables also yields desirable statistical properties as we will show in Chapter 3.

What Interactions Signify in Regression

In multiple regression analysis, the relationship of each predictor to the criterion is measured by the slope of the regression line of the criterion Y on the predictor; the regression coefficients are these slopes. Consider first the familiar two predictor regression equation $\hat{Y} = b_1X + b_2Z + b_0$, which contains no interaction. In this equation, the slope (or regression) of Y on X has a constant value across the range of Z. If one calculated the regression of Y on X for all cases at any single value of Z, the regression coefficient for X in the subsample of cases would equal b_1 in the overall equation. Put another way, the regression of Y on X is independent of Z. The regression of Y on Z is represented by b_2, which has a constant value across the range of X in this regression equation.

Now consider equation 2.1, which contains an XZ interaction. The XZ interaction signifies that the regression of Y on X depends upon the specific value of Z at which the slope of Y on X is measured. There is a different line for the regression of Y on X at each and every value of Z. The regressions of Y on X at specific values of Z form a family of regression lines. Each of these regression lines at one value of Z is referred to as a *simple regression line*. Because the regression of Y on X depends upon the value of Z, the effect of X in equation 2.1 is termed a *conditional* effect (e.g., Darlington, 1990).

The XZ interaction is symmetrical. The presence of the XZ interaction in equation 2.1 equivalently means that the effect of predictor Z is *conditional* on X; there is a different regression of Y on Z at each value of X.

Data Set for Numerical Examples

To illustrate our prescriptions for post hoc probing of interactions, a single data set is employed thoughout this and the next chapter. In our example we will predict the self-assurance of managers (criterion Y) based on two predictors, their length of time in the managerial position (X) and their managerial ability (Z). The data we use are artificial: They were specifically constructed to include an interaction between X and Z. In these data, individuals high in managerial ability increase in their self-assurance with increased time in the position, whereas individuals low in managerial ability decrease in self-assurance with increased time in the position. As we have described the expected relationships, the regression of Y on X varies as a function of Z. All three variables, self-assurance (Y), time in position (X), and managerial ability (Z) are continuous variables.

Our simulation is based on moderately correlated bivariate normal predictors X and Z and their interaction XZ; 400 cases were generated in all. From these two predictors, the XZ cross-product term was formed. Predicted scores were generated from X, Z, and the interaction. Random error was added to the predicted scores to create observed criterion scores. Finally, the regression equation, $\hat{Y} = b_1X + b_2Z + b_3XZ + b_0$, was estimated based on the three original predictors and the observed scores (Y).

The results of the regression analysis and the post hoc probing of the XZ interaction that we will be discussing in this chapter are presented in Tables 2.1, 2.2, and 2.3 and Figure 2.1. Note that in this chapter we will only discuss the portions of the tables that pertain to "centered data;" the portions of the tables labeled "uncentered data" are discussed in Chapter 3. Of most importance at present, we see in Table 2.1c(ii) the regression equation containing the interaction:

$$\hat{Y} = 1.14X + 3.58Z + 2.58XZ + 2.54$$

In this equation, both the Z effect and the XZ interaction are significant. From the overall equation it appears that there is no overall effect of length of time in position (X) on self-assurance (Y), that there is a positive effect of managerial ability on self-assurance (Z), but that the relationship of time to self-assurance is modified by managerial ability (XZ). Also of interest, Table 2.1a presents the means and standard deviations for X and Z and the correlation matrix for X, Z, and XZ for the centered data. Table

Table 2.1
Centered versus Uncentered Regression Analyses Containing an Interaction

a. Centered Data
$\bar{X} = 0 (s_X = 0.95)$
$\bar{Z} = 0 (s_Z = 2.20)$

Correlation matrix

	X	Z	XZ	Y
X	—	.42	.10	.17
Z		—	.04	.31
XZ			—	.21

b. Uncentered Data
$\bar{X}' = 5 \ (s'_X = 0.95)$
$\bar{Z}' = 10 (s'_Z = 2.20)$

Correlation matrix

	X'	Z'	$X'Z'$	Y
X'	—	.42	.81	.17
Z'		—	.86	.31
$X'Z'$			—	.21

c. Regression Equations Based on Centered Data
(i) No interaction: $\hat{Y} = 1.67X + 3.59Z^{**} + 4.76$

(ii) With interaction: $Y = 1.14X + 3.58Z^{**} + 2.58XZ^{**} + 2.54$

d. Regression Equations Based on Uncentered Data
(i) No interaction: $\hat{Y} = 1.67X' + 3.59Z'^{**} - 39.47$

(ii) With interaction: $Y = -24.68X'^{**} - 9.33Z'^{**} + 2.58X'Z'^{**} + 90.15$

$^{**}p < .01$; $^{*}p < .05$

2.1c(i) shows the regression equation when the interaction term is omitted.[1]

Probing Significant Interactions in Regression Equations

Given that a significant interaction has been obtained, we now wish to probe this interaction to sharpen our understanding of its meaning. The primary techniques for probing of this term are plotting the interaction and post hoc statistical testing.

Plotting the Interaction

Probing a significant interaction in MR begins with recasting the regression equation as the regression of the criterion on one predictor.[2] For example, the regression equation is restructured through simple algebra to express the regression of Y on X at levels of Z:

$$\hat{Y} = (b_1 + b_3 Z)X + (b_2 Z + b_0) \quad (2.2)$$

In this restructured form of equation 2.1, the slope of the regression of Y on X, $(b_1 + b_3 Z)$, depends upon the particular value of Z at which the slope is considered. We refer to $(b_1 + b_3 Z)$ as the *simple slope* of the regression of Y on X at Z. By *simple slope* we mean the slope of the regression of Y on X at (conditional on) a single value of Z. Note that the simple slope $(b_1 + b_3 Z)$ combines the regression coefficient of Y on X (b_1) with the interaction coefficient (b_3). Readers familiar with ANOVA may find it helpful to think of simple slopes as the analog in MR of simple effects in ANOVA.

We must then choose several values of Z to substitute into equation 2.2 to generate a series of *simple regression equations*. If Z were categorical, as for example when Z represents a dichotomous variable such as gender, then we would compute two simple regression equations, one for men and one for women. (Chapter 7 provides an in-depth consideration of regression involving categorical and continuous variables.) On the other hand, if Z is continuous, as in the example of managerial ability, then the investigators are free to choose any value within the full range of Z. In some cases, theory, measurement considerations, or previous research may suggest interesting values of Z that should be chosen. For example, in a clinical diagnostic test, if a specific score represented a cutoff above which pathology were indicated, then that cutoff score, a higher score typical of the clinical condition, and a lower score typical of normal populations might be chosen. In a study involving income, the federal government's value of the poverty line for a family of four might be chosen. In other cases, such as our fictitious example of managerial ability, no social science based rationale will exist to guide the choice of several values of Z. In such cases, Cohen and Cohen (1983) have suggested as a guideline that researchers use the values Z_M, Z_H, Z_L, corresponding to the mean of Z, one standard deviation above $\bar{Z}$, and one standard deviation below $\bar{Z}$, respectively. Whatever values of Z are chosen, each is substituted into equation 2.2 to generate a series of simple regression equations of Y on X at specific values of Z. These equations are plotted to display the interaction.

Numerical Example

Figure 2.1a depicts three simple regression lines of the regression of self-assurance (Y) on time in position (X) as a function of three values of managerial ability, Z_L, Z_M, and Z_H for our data set. Note the symmetrical pattern of the three simple regression lines that is characteristic of equations with a significant XZ interaction term.

To generate these simple regression lines, the overall regression equation $\hat{Y} = 1.14X + 3.58Z + 2.58XZ + 2.54$ was rearranged to show the regression of Y on X at levels of Z:

$$\hat{Y} = (1.14 + 2.58Z)X + (3.58Z + 2.54) \qquad (2.3)$$

Then, following Cohen and Cohen (1983), values of Z were chosen to be one standard deviation below the mean ($Z_L = -2.20$), at the mean ($Z_M = 0$), and one standard deviation above the mean ($Z_H = 2.20$). Simple regression lines were then generated by substituting these values (-2.20, 0, 2.20) in turn into equation 2.3. For example, to generate the simple regression for $Z_H = 2.20$, the following substitution was made:

$$\begin{aligned}\hat{Y} &= [1.14 + 2.58(2.20)]X + [3.58(2.20) + 2.54]\\ &= 6.82X + 10.41.\end{aligned}$$

The results of the computations of simple regression equations for Z_M and Z_L are given in Table 2.2a. The simple regression equations indicate a positive regression of Y on X for Z_H, a negative regression of Y on X for

Table 2.2
Simple Regression Equations for Centered and Uncentered Data

a. Regression of Y on X at Particular Values of Z for Centered Data
In general: $\hat{Y} = (1.14 + 2.58Z)X' + (3.58Z + 2.54)$

At $Z_H = 2.20$: $\hat{Y} = 6.82X + 10.41$

At $Z_M = 0.00$: $\hat{Y} = 1.14X + 2.54$

At $Z_L = -2.20$: $\hat{Y} = -4.54X - 5.33$

b. Regression of Y on X' at Particular Values of Z' for Uncentered Data
In general: $\hat{Y} = (-24.68 + 2.58Z')X' + (-9.33Z' + 90.15)$

At $Z_{H'} = 12.20$: $\hat{Y} = 6.82X' - 23.67$

At $Z_{M'} = 10.00$: $\hat{Y} = 1.14X' - 3.15$

At $Z_{L'} = 7.80$: $\hat{Y} = -4.54X' + 17.38$

NOTE: Regression equation rearranged to show regression of Y on X at levels of Z: $\hat{Y} = (b_1 + b_3Z)X + (b_2Z + b_0)$

Z_L, and essentially no relationship between X and Y for Z_M. Figure 2.1a reveals a complex pattern of regression of Y on X depending on the level of Z. If only the nonsignificant b_1 coefficient in Table 2.1c(ii) had been examined, it would have been concluded that there was no relationship of X to Y.

Post Hoc Probing

Once plotting is accomplished, two questions that parallel the probing of ANOVA interactions with simple effects may be asked: (a) For a specified value of Z, is the regression of Y on X significantly different from zero, and (b) for any pair of simple regression equations, do their slopes differ from one another? (Similar questions may also be asked for the regression of Y on Z for each value of X.)

Is the Slope of the Simple Regression Line Significantly Different From 0?

The approach to probing interactions prescribed here permits the testing of the significance of the simple slopes of regression lines at single values of a second predictor. The approach was described in Friedrich (1982; see also Darlington, 1990; Jaccard, Turrisi, & Wan, 1990) for the case

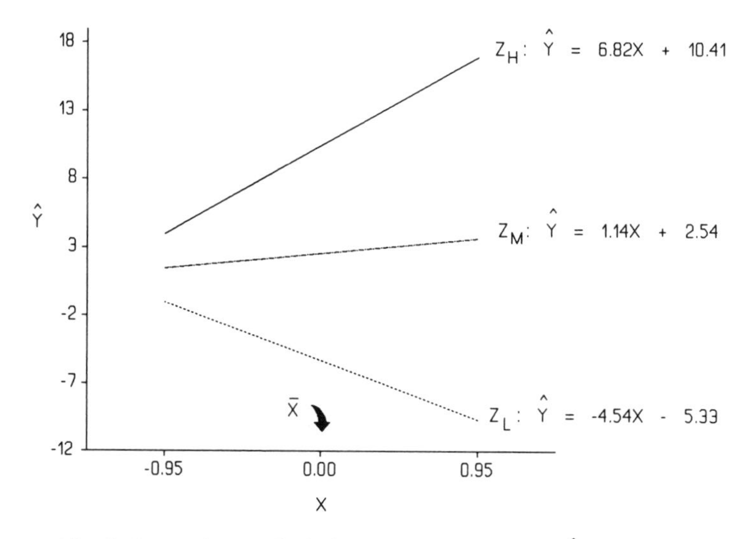

a. Simple Regression Analysis from Centered Analysis: $\hat{Y} = 1.14X + 3.58Z + 2.58XZ + 2.54$

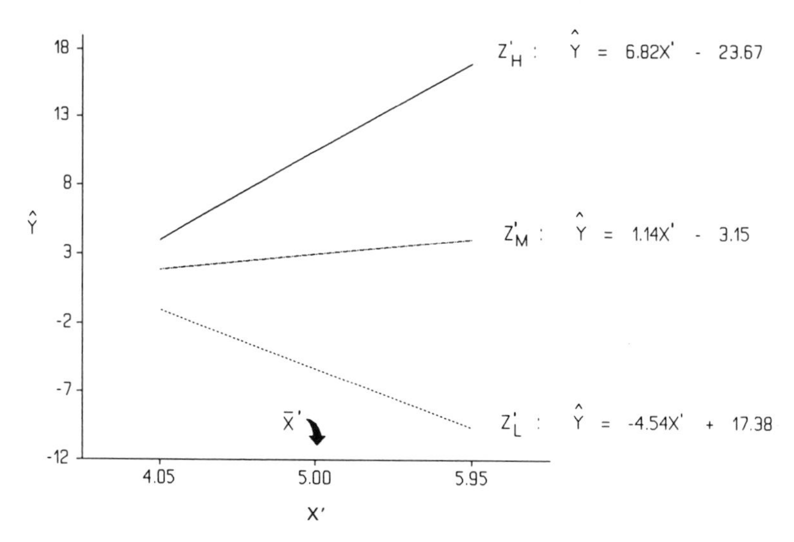

b. Simple Regression Equation from the Uncentered Analysis: $\hat{Y} = -24.68X' - 9.33Z' + 2.58X'Z' + 90.15$

Figure 2.1. Interaction Plotted from Centered and Uncentered Equations

of the XZ interaction. It involves the calculation of the standard errors of the simple slopes of simple regression equations. Then t-tests for the significance of the simple slopes are computed. We initially provide an overview and numerical example of this strategy. We then show how these tests of simple slopes can easily be accomplished with available regression analysis computer programs. An optional section is presented at the end of this chapter for more advanced readers who wish a general derivation of standard errors of simple slopes.

To calculate the standard error of the simple slope we use values from the variance–covariance matrix of the regression coefficients.[3] Estimates of these values are produced by regression programs in standard statistical packages such as SPSS-X and SAS. Specific elements from this matrix corresponding to terms in the simple slope are given a weight and then combined to produce the estimate of the standard error of the simple regression coefficient.

Returning to the simple slope $(b_1 + b_3 Z)$ in equation 2.2, its standard error is given as

$$s_b = \sqrt{s_{11} + 2Zs_{13} + Z^2 s_{33}} \tag{2.4}$$

The values s_{11} and s_{33} are the variances of b_1 and b_3, respectively, taken from $\mathbf{S}_b$, the sample estimate of the variance-covariance matrix of predictors; s_{13} is the covariance between b_1 and b_3 taken from $\mathbf{S}_b$. As the value of Z varies in the simple slope, the value of the standard error in equation 2.4 varies as well. Note that equation 2.4 pertains *only* to the simple slope $(b_1 + b_3 Z)$ in equation 2.2.

t-Tests for Simple Slopes. The t-test for whether a simple slope differs from zero is simply the value of the simple slope divided by its standard error with $(n - k - 1)$ degrees of freedom, where n is the number of cases and k is the number of predictors, not including the regression constant (here $k = 3$).

Numerical Example. The covariance matrix of regression coefficients $\mathbf{S}_b$ is given in Table 2.3a for the simulated data set with centered predictors. This matrix was obtained from the regression program of the SPSS-X package applied to the centered data set[4]. Values from this matrix are used in computing the standard errors of the simple slopes. Recall that in Figure 2.1a and Table 2.2a, the simple slopes for the regression of Y on X at Z_H, Z_M, and Z_L were 6.82, 1.14, and −4.54, respectively. Equation 2.4 yields the standard errors associated with each slope. The equation involves three elements from $\mathbf{S}_b$, specifically $s_{11} = 2.35$, $s_{13} = -0.08$, and $s_{33} = 0.40$. Substituting $Z_H = 2.20$ into equation 2.4 yields the standard error of the simple slope of Y on X at Z_H:

$$s_H = \sqrt{2.35 + 2(2.20)(-0.08) + (2.20^2)(0.40)} = \sqrt{3.93} = 1.98$$

Similar substitutions of $Z_M = 0$, and $Z_L = -2.20$ yield the estimated standard errors $s_M = 1.53$ and $s_L = 2.15$, also summarized in Table 2.3a. Finally, in Table 2.3a, the t-tests of each simple slope against zero are

Table 2.3
Computation of Standard Errors and t-Tests for Simple Slopes

a. Analysis of Centered Data
 (i) Covariance matrix of regression coefficients

$$\mathbf{S}_b = \begin{array}{c} b_1 \\ b_2 \\ b_3 \end{array} \overset{\begin{array}{ccc} b_1 & b_2 & b_3 \end{array}}{\begin{bmatrix} 2.35 & -0.41 & -0.08 \\ -0.41 & 0.43 & -0.00 \\ -0.08 & -0.00 & 0.40 \end{bmatrix}}$$

 (ii) Simple slopes, standard errors of simple slopes, and t-tests

Simple slope	Standard error	t-test
$b_H = 6.82$	$s_H = 1.98$	$t = 6.82/1.98 = 3.45^{**}$
$b_M = 1.14$	$s_M = 1.53$	$t = 1.14/1.53 = 0.74$
$b_L = -4.54$	$s_L = 2.15$	$t = -4.54/2.15 = -2.11^{*}$

b. Analysis of Uncentered Data
 (i) Covariance matrix of regression coefficients

$$\mathbf{S}_{b'} = \begin{array}{c} b'_1 \\ b'_2 \\ b'_3 \end{array} \overset{\begin{array}{ccc} b'_1 & b'_2 & b'_3 \end{array}}{\begin{bmatrix} 43.88 & 19.96 & -4.07 \\ 19.96 & 10.42 & -2.00 \\ -4.07 & -2.00 & 0.40 \end{bmatrix}}$$

 (ii) Simple slopes, standard errors of simple slopes and t tests

Simple slope	Standard error	t-test
$b_{H'} = 6.82$	$s_{H'} = 1.98$	$t = 6.82/1.98 = 3.45^{**}$
$b_{M'} = 1.14$	$s_{M'} = 1.53$	$t = 1.14/1.53 = 0.74$
$b_{L'} = -4.54$	$s_{L'} = 2.15$	$t = -4.54/2.15 = -2.11^{*}$

$^{**}p < .01$; $^{*}p < .05$

provided. These tests confirm the positive regression Y on X at Z_H and the negative regression of Y on X at Z_L; the regression of Y on X at Z_M does not differ from 0.

The Regression of Y on Z. The above presentation and numerical example pertain to the regression of Y on X at levels of Z. If instead we were interested in the regression of Y on Z at levels of X, the simple slope equation would be expressed as $\hat{Y} = (b_2 + b_3X)Z + (b_1X + b_0)$. For this equation the simple slope is $(b_2 + b_3X)$ and its standard error is

$$s_b = \sqrt{s_{22} + 2Xs_{23} + X^2 s_{33}} \qquad (2.5)$$

Once again, the t-test for whether a simple slope differs from zero is simply the value of the simple slope divided by its standard error with $(n - k - 1)$ degrees of freedom.

Simple Slope Analysis by Computer. A computer procedure using standard regression programs can be used to perform the entire simple slope analysis (Darlington, 1990, personal communication 1990; Judd & McClelland, 1989). Here we confine the presentation to the regression of Y on X at values of Z in equation 2.2. The procedure generalizes to more complex equations. Examples of its use are presented throughout the text.

Following Darlington (1990), we use the term *conditional value of Z* (CV_Z) to refer to the specific value of Z at which the regression of Y on X is considered. Suppose we seek the simple slope for the regression of Y on X at $Z_H = 2.20$, one standard deviation above the mean of Z; then $CV_Z = 2.20$. To carry out the simple slope analysis by computer, predictor Z is transformed to a new variable Z_{CV} by subtracting CV_Z from Z (i.e., $Z_{CV} = Z - CV_Z$). Then Z_{CV} is used in the regression analysis in lieu of Z. The value of b_1 from that analysis is the simple slope of the regression of Y on X at $Z = 2.20$.

In sum, three steps are required to carry out the simple regression analysis by computer:

1. Create a new variable Z_{CV}, which is the original variable Z minus the conditional value of interest, that is, $Z_{CV} = Z - CV_Z$;
2. form the crossproduct of the new variable with predictor X, that is, $(X)(Z_{CV})$; and
3. regress the criterion Y on X, Z_{CV}, and $(X)(Z_{CV})$.

The resulting regression coefficient b_1 will be the desired simple regression coefficient of Y on X at the conditional value CV_Z of Z. The regression constant (intercept) from that analysis will be that for the simple regression equation. The standard error of b_1 will be the standard error of the simple slope of Y on X at CV_Z, and the t-test will be that for the simple slope.

Similarly, if the simple slope of Y on X one standard deviation *below* the mean of centered Z is sought (here $CV_Z = -2.20$), then once again a new variable $Z_{CV} = Z - CV_Z$ is calculated, here $Z_{CV} = Z - (-2.20)$ and the regression of Y on X, Z_{CV}, and $(X)(Z_{CV})$ is performed. The resulting b_1 term, its standard error, and t-test form the simple slope analysis one standard deviation *below* the mean of Z, at $CV_Z = -2.20$.

Table 2.4 provides SPSSX computer output for the simple slope analysis reported in Table 2.3a. First, the overall regression analysis with centered X, Z, and XZ is given, replicating Table 2.1c(ii), and two transformed variables are calculated:

1. ZABOVE = Z − (2.20), for the regression of Y on X at $CV_Z = 2.20$, one standard deviation above the mean of Z, and
2. ZBELOW = Z − (−2.20), for the regression of Y on X at $CV_Z = -2.20$, one standard deviation below the mean of Z.

Second, their crossproducts with X are calculated: XZABOVE and XZBELOW. Third, the two regression analyses described above are performed. In Table 2.4c, the regression analysis involving X, ZBELOW, and XZBELOW is reported. The regression coefficient b_1 and constant b_0 equal those for the simple regression equation of Y on X at Z_L in Table 2.3a; the standard error and t-test are identical as well. In Table 2.4d, the regression analysis involving X, ZABOVE, and XZABOVE is reported and corresponds to the simple regression of Y on X at Z_H in Table 2.3a.

This computer approach to simple slope analysis generalizes to more complex regression analyses, for example, those involving three linear predictors and their interactions and those involving higher order polynomial terms (e.g., X^2, X^2Z), considered in Chapters 4 and 5, respectively.

Do the Slopes of a Pair of Simple Regression Lines Differ From One Another?

Having determined for which values of Z the regression of Y on X is different from zero, the investigator may wish to determine whether the simple slope of Y on X differs at two values of Z, say Z_H versus Z_L, as previously defined. The simple slopes in question are $(b_1 + b_3Z_H)$ versus $(b_1 + b_3Z_L)$, and their difference is simply a function of b_3, that is, $d =$

$(b_3 Z_H - b_3 Z_L) = (Z_H - Z_L)\, b_3$. The t-test of the difference between the slopes is given by

$$t = \frac{d}{s_d} = \frac{(Z_H - Z_L)\, b_3}{\sqrt{(Z_H - Z_L)^2 s_{33}}} = \frac{b_3}{\sqrt{s_{33}}}$$

Note that this is identical to the t-test for the significance of the b_3 coefficient in the overall analysis. In other words, given that Z is a continuous variable, the significance of the b_3 coefficient in the overall analysis in-

Table 2.4
Computation of Simple Slope Analysis by Computer for the XZ Interaction in the Regression Equation $\hat{Y} = b_1 X + b_2 Z + b_3 XZ + b_0$

a. Overall Analysis with Centered X and Centered Z

(i) Means and standard deviations

	Mean	Std Dev
Y	4.759	28.019
X	0.000	0.945
Z	0.000	2.200
XZ	0.861	2.086

(ii) Variance–covariance matrix of regression coefficients (b)
Below diagonal: covariance; above: correlation

	X	Z	XZ
X	2.34525	−0.41324	−0.08489
Z	−0.41469	0.42938	−0.00453
XZ	−0.08211	−0.00187	0.39895

(iii) Regression Analysis

Variable	B	SE B	T	Sig T
X	1.136404	1.531420	0.742	.4585
Z	3.577193	0.655271	5.459	.0000
XZ	2.581445	0.631627	4.087	.0001
(Constant)	2.537403	1.418212	1.789	.0744

b. Computation of ZABOVE, ZBELOW, and crossproduct terms required for simple slope analysis

```
COMPUTE ZABOVE = Z - 2.20
COMPUTE ZBELOW=Z-(-2.20)
COMPUTE XZABOVE=X*ZABOVE
COMPUTE XZBELOW=X*ZBELOW
```

(Table 2.4, continued)

c. Regression Analysis with ZBELOW AND XZBELOW, Yielding Simple Slope Analysis at Z_L (Regression of *Y* on *X* One Standard Deviation Below the Mean of *Z*)

(i) Means and standard deviations

	Mean	Std Dev
Y	4.759	28.019
X	0.000	0.945
ZBELOW	2.200	2.200
XZBELOW	0.861	3.082

(ii) Regression Analysis

Variable	B	SE B	T	Sig T
X	−4.542777	2.153481	−2.110	.0355
ZBELOW	3.577193	0.655271	5.459	.0000
XZBELOW	2.581446	0.631627	4.087	.0001
(Constant)	−5.332421	2.020502	−2.639	.0086

d. Regression Analysis with ZABOVE AND XZABOVE, Yielding Simple Slope Analysis at Z_H (Regression of *Y* on *X* One Standard Deviation Above the Mean of *Z*)

(i) Means and standard deviations

	Mean	Std Dev
Y	4.759	28.019
X	0.000	.945
ZABOVE	−2.200	2.200
XZABOVE	0.861	2.801

(ii) Regression Analysis

Variable	B	SE B	T	Sig T
X	6.815584	1.978606	3.445	.0006
ZABOVE	3.577193	0.655271	5.459	.0000
XZABOVE	2.581446	0.631627	4.087	.0001
(Constant)	10.407229	2.024012	5.142	.0000

dicates that the regression of *Y* on *X* varies across the range of *Z*. No further test is required of whether simple slopes of *Y* on *X* differ from one another as a function of the value of *Z*.

A Caution Concerning the Use of Simple Slope Tests

When simple slopes are evaluated using a priori values or population-based values as in the example of the clinical diagnostic test described above, the procedures described in this chapter provide proper values of the test statistics. Similarly, when a high or low value of Z is chosen on the basis of Cohen and Cohen's (1983) or other guidelines and the researcher's interest is in making a statement about the simple slope at that specific numeric value, the test statistics are unbiased. These cases are the ones most likely to be encountered by researchers.

In contrast, if the researcher's interest centers on making inferences about the simple slope at a specific population-based value (e.g., the population mean; one standard deviation above the mean in the population), the t-tests of simple slopes described in this and subsequent chapters are positively biased. The magnitude of this bias decreases with increasing sample size. Two remedies for this problem have been summarized by West and Aiken (1990): (a) A procedure developed by Lane (1981) to provide conservative tests of the value of the simple slopes may be used. (b) Bootstrapping techniques (Darlington, 1990, personal communication 1990; Stine, 1990) may be used to provide empirical estimates of the standard error. At the present time we know of no systematic investigation of the extent of bias and the adequacy of the proposed remedies.

Ordinal Versus Disordinal Interactions

A useful distinction, borrowed from the ANOVA literature, is the classification of interactions as being *disordinal* versus *ordinal* (Lubin, 1961), or, equivalently, *crossover* versus *noncrossover*, respectively. According to this descriptive classification, the interaction is ordinal (noncrossover) when the simple regression lines (or lines representing levels of one factor with categorical variables) for an interaction do not cross within the *possible range* of the values of the other variable. Conversely, the interaction is disordinal (crossover) when the simple regression lines cross within the possible range of values of the other variable. For example, if the scale measuring managerial ability in our example had a potential range of values from 1 to 7 and the crossing point were at 5, the interaction would be disordinal (crossover). In contrast, if the lines were to cross at -2 or at 12, the interaction would be ordinal.

Potential difficulties with this descriptive classification can develop in the MR context with interactions between continuous variables that have no obvious extreme points that define the ends of the continuum. One approach to such cases is for the researcher to examine the calculated crossing point relative to the *actual range* of the data. Interactions whose crossing point falls outside the actual range of values on X are classified as being ordinal, whereas those whose crossing point falls inside the actual range of values on X are classified as being disordinal. An alternative approach in the absence of scale-based or data-based criteria is for the researcher to consider a *meaningful range* of the variable in terms of the system being characterized by the regression equation; this meaningful range has been referred to as the *dynamic range* of the variable in the context of sensory systems such as vision or audition (see Teghtsoonian, 1971). Those interactions for which the lines crossed within the meaningful range of the variable would be termed disordinal, whereas other interactions whose crossing point fell outside this range would be termed ordinal.

The reader should bear in mind that the classification of an interaction as ordinal versus disordinal is always with regard to a particular configuration of variables. An interaction may be ordinal in one direction, say the regression of Y on X at values of Z, and disordinal in the other direction (Y on Z at values of X). The question of whether to characterize an interaction in terms of Y and X at values of Z or in terms of Y and Z at values of X may be driven by theory and the specific predictions to be tested. For example, most theoretical discussions present life stress (X) as the predictor of health (Y), with social support (Z) being described as the variable that moderates this relationship; hence the regression of Y on X at values of Z is considered. However, in general, it is useful to examine both the regression of Y on X at levels of Z, and Y on Z at levels of X. Both castings provide potentially informative and complementary two-dimensional representations of what is in reality a three-dimensional regression surface.

Determining the Crossing Point in an Interaction

The point at which two simple regression lines cross can be determined algebraically. For the regression of Y on X at values of Z, the simple regression equation is written at two specific values of Z, say Z_{H} and Z_{L}, yielding two simple regression equations:

$$\hat{Y}_{\mathrm{H}} = (b_1 + b_3 Z_{\mathrm{H}})X + (b_2 Z_{\mathrm{H}} + b_0)$$

$$\hat{Y}_{\mathrm{L}} = (b_1 + b_3 Z_{\mathrm{L}})X + (b_2 Z_{\mathrm{L}} + b_0)$$

The two equations are set equal to one another to determine the expression for the point at which the lines represented by these interactions cross.

Here,

$$X_{\text{cross}} = \frac{-b_2}{b_3} \tag{2.6}$$

Note that if there is no interaction (i.e., $b_3 = 0$), the simple regression lines do not cross.

The crossing point for the simple regression of Y on Z at values of X can be derived in a parallel manner. This crossing point is $Z_{\text{cross}} = -b_1/b_3$.

Numerical Example

For the numerical example in Table 2.1c(ii) with $b_2 = 3.58$ and $b_3 = 2.58$, the simple regressions of Y on X at values of Z cross at the value $X_{\text{cross}} = -3.58/2.58 = -1.39$. For centered X ($\overline{X} = 0.0$ and $s_X = 0.95$) the simple regression lines cross $(-1.39 - 0.0)/0.95 = -1.47$ standard deviations below the mean. Figure 2.1 portrays the interaction, but with the range of X limited to one standard deviation on either side of its mean. The interaction is ordinal within this range. If this is the meaningful range of X, then we recommend classifying the interaction as ordinal and explaining the limits of X considered in so doing.

Optional Section: The Derivation of Standard Errors of Simple Slopes

In this section we provide the general derivation of standard errors of any simple slope in any ordinary least squares (OLS) regression equation of the form

$$\hat{Y} = b_1X + b_2Z + \cdots + b_pW + b_0$$

that is, any regression equation that is linear in the regression coefficients (Kmenta, 1986). The approach pertains to all of the more complex regression equations in this book. Readers familiar with matrix algebra should find the exposition straightforward. For readers unfamiliar with matrix algebra, the general form of the expression for the variance of a simple slope (square of the standard error) is given in equation 2.10 below.

The starting point for deriving standard errors of simple slopes is the observation that each simple slope is a linear combination of the original regression coefficients in the equation. In the equation $\hat{Y} = (b_1 + b_3Z)X + (b_0 + b_2Z)$, the simple slope for the regression of Y on X is

$(b_1 + b_3Z)$. Using the known properties of linear combinations, we can derive the sampling variance of the simple slope $(b_1 + b_3Z)$.

Consider any linear combination U of variables $b_1 \cdots b_p$, weighted by $w_1 \cdots w_p$, respectively. In vector equation form, this may be expressed as $U = \mathbf{w}'\mathbf{b}$, or equivalently in algebraic form

$$U = w_1b_1 + w_2b_2 + \cdots + w_pb_p.$$

Here the regression coefficents $\mathbf{b}' = [b_1\ b_2 \cdots b_p]$ are the elements of the combination and $\mathbf{w}' = [w_1\ w_2 \cdots w_p]$ are the weights that define the combination. The variance σ_b^2 of the combination is a function of Σ_b, the variance covariance matrix of the elements $b_1 \cdots b_p$ and of the weights themselves, as given by the quadratic form

$$\sigma_b^2 = \mathbf{w}'\Sigma_b\mathbf{w} \tag{2.7}$$

As already explained, in the case of usual ordinary least squares regression analysis, the variance–covariance matrix required is that of the regression coefficients themselves, $\mathbf{S}_b$, the sample estimate of Σ_b. Under the typical assumption of ordinary least squares (OLS) regression, namely normally distributed residuals ϵ_i with mean zero $[\mathrm{E}(\epsilon_i = 0)]$ and variance σ_ϵ^2, the least squares estimates of the regression coefficients are normally distributed, with the estimate of their variance-covariance matrix[5] given by the matrix equation $\mathbf{S}_b = \mathrm{MS}_{Y-\hat{Y}}\mathbf{S}_{XX}^{-1}$. In this equation, $\mathrm{MS}_{Y-\hat{Y}}$ is the mean square residual from the overall analysis of regression (ANOReg), and $\mathbf{S}_{XX}^{-1}$ is the inverse of the covariance matrix of predictors. (See Maddala, 1977, for a clear exposition of the sampling properties of OLS estimators in multiple regression analysis, and Morrison, 1976, for comments on the properties of linear combinations of normally distributed variables.)

The simple slope is written as a linear combination of all the coefficients in the equation except the constant b_0. For $(b_1 + b_3Z)$, it is rewritten as $U = (1)b_1 + (0)b_2 + (Z)b_3$, with the weight vector $\mathbf{w}' = [1\ 0\ Z]$. At any particular value of Z, the sample estimate of the variance of the simple slope $(b_1 + b_3Z)$ is then given as

$$\mathbf{S}_b^2 = \mathbf{w}'\mathbf{S}_b\mathbf{w} = [1\ 0\ Z]\begin{bmatrix} s_{11} & s_{12} & s_{13} \\ s_{21} & s_{22} & s_{23} \\ s_{31} & s_{32} & s_{33} \end{bmatrix}\begin{bmatrix} 1 \\ 0 \\ Z \end{bmatrix} \tag{2.8}$$

In this equation s_{jj} is the variance of the estimate of regression coefficient b_j, and s_{ij} is the covariance between the estimates of b_i and b_j. Completing the multiplication yields the following expression for the variance of the simple slope of Y on X at values of Z in equation 2.2. This is the square of the standard error given in equation 2.4.

$$s_b^2 = s_{11} + 2Zs_{13} + Z^2 s_{33} \tag{2.9}$$

For the regression of Y on Z at values of X, the simple slope is $(b_2 + b_3X)$, and hence $\mathbf{w}' = [0\ 1\ X]$. With this weight vector used in place of the vector $[1\ 0\ X]$ the variance of the simple slope of Y on Z at values of X is generated; it is the expression under the radical in equation 2.5.

Equation 2.7, the general equation for the variance of any linear combination of regression coefficients, may also be expressed in algebraic form (Stolzenberg & Land, 1983, p. 657). In terms of population values, the general expression is as follows:

$$\sigma_b^2 = \sum_{j=1}^{k} w_j^2 \sigma_{jj} + \sum_{j=1}^{k} \sum_{i \neq j} w_i w_j \sigma_{ij} \tag{2.10}$$

where

k is the total number of regression coefficients for predictors excluding b_0 in the equation (for equation 2.1, $k = 3$);

w_j are the weights used to define the combination in the weight vectors defined above, and w_j^2 are their squares;

σ_{jj} is the variance of regression coefficient b_j of which s_{jj} is the sample estimate, as in equation 2.5 above;

σ_{ij} is the covariance between two regression coefficients b_i and b_j, of which s_{ij} is the sample estimate, as in equation 2.5 above.

Throughout Chapters 4, 5, and 7 we state equations for the variances of simple slopes in a variety of regression equations. All the equations follow the form of equation 2.10. As we work through various regression equations, we will indicate the particular weight vectors involved in deriving the simple slope variances. The reader may then follow equation 2.7 in matrix form or, equivalently, 2.10 in algebraic form, to confirm the formulas for the simple slope variances.

Summary

This chapter addressed the probing of significant interactions between two continuous variables X and Z in the regression equation $\hat{Y} = b_1X + b_2Z + b_3XZ + b_0$. First, the regression equation was rearranged to show the regression of the criterion on X at values of Z; the simple slope of that regression equation was defined. Post hoc probing of the interaction began with prescriptions for plotting the interaction. A t-test for the significance of the simple slopes was presented together with a simple, computer-based method for performing this test. The distinction between ordinal and disordinal (noncrossover versus crossover) interactions was presented for the interaction between two continuous variables, and the procedure for determining the crossing point of simple regression lines was illustrated. Finally, a more advanced optional section presented a general derivation of the standard errors of simple slopes in any OLS regression equation.

Notes

1. In practice there will typically be little difference between the b_1 and b_2 coefficients in the regression equations containing the interaction [e.g., Table 2.1c(ii)] and those coefficients in the regression equation not containing the interaction [e.g., Table 2.1c(i)] if predictors X and Z are centered and have an approximately bivariate normal distribution.

2. In much of the discussion we refer to the regression of Y on X at values of Z. The interaction may just as well be cast in terms of the regression of Y on Z at values of X. Our exposition is often confined to Y on X and values of Z for simplicity.

3. Conceptually, the population variance-covariance matrix of the regression coefficients, Σ_b, can be understood as follows. Imagine computing equation 2.1 for a infinite number of random samples from a given population. The variance of each regression coefficient (e.g., b_1) across all the samples would be on the main diagonal of Σ_b. The covariance between pairs of regression coefficients (e.g., b_1 with b_2) across all samples would be the off-diagonal entries.

4. In SPSS-X the covariances among estimates (s_{12}, s_{13}, s_{23}) and the correlations among the estimates are printed in the same matrix, with variances of the estimates (s_{11}, s_{22}, s_{33}) on the main diagonal, covariances below the diagonal, and correlations above the diagonal. This matrix is obtained in SPSS-X REGRESSION with the BCOV keyword on the STATISTICS subcommand. To form the covariance matrix $\mathbf{S}_b$ in Table 2.3a(i), we have placed the covariances both below and above the diagonal; SPSS users should be certain to do this as well.

The covariance matrix of the estimates is obtained in SAS from PROC REG with the keyword COVB on the MODEL statement. The $\mathbf{S}_b$ matrix obtained in SAS contains only variances and covariances, just as in Table 2.3a(i). No modification of the SAS output is required.

5. In general, the covariance matrix of the parameters is the inverse of Fisher information matrix (Rao, 1973).

3 The Effects of Predictor Scaling on Coefficients of Regression Equations

In Chapter 1, we introduced the problem of the lack of invariance of regression coefficients in equations containing interactions even under simple linear transformations of the data. However, to this point, we have not specifically addressed this problem, which has led to considerable confusion in the literature (see discussions by Friedrich, 1982; Schmidt, 1973; Sockloff, 1976). In this chapter we explore the problem both algebraically and by numerical example for the case of the regression equation containing one XZ interaction term. After the consideration of scaling effects to provide the reader with the necessary understanding of centered versus uncentered solutions, we then examine the interpretation of each of the regression coefficients in equation 2.1, $\hat{Y} = b_1X + b_2Z + b_3XZ + b_0$. Finally we explore the relationship between the centered solution and several potential standardized solutions, showing that only the procedure proposed by Friedrich (1982) produces a fully interpretable standardized solution.

The Problem of Scale Invariance

The problem of lack of scale invariance under linear transformation of predictors is shown algebraically in this section. The scale transformation we consider is rescaling by *additive constants* (i.e., adding or subtracting constants from predictor scores). For example, suppose we have a raw score X'; we subtract the mean (a constant) from each score, yielding the centered score X. Or, we rescale a variable that ranges from -3 to $+3$ by adding 4 to each score, so that resulting scores will range from 1 to 7. In our usual experience such rescaling has no effect on the correlational properties of the rescaled variables, and hence will have no effect on linear regression. This is the desired state of affairs: We want the solution to be identical for the original raw and transformed variables. However, when there are interactions in the regression equation, simply rescaling by additive constants has a profound effect on regression coefficients.

We examine rescaling effects using the regression equation presented in Chapter 2:

$$\hat{Y} = b_1X + b_2Z + b_3XZ + b_0 \tag{3.1}$$

or as rewritten to show the regression of Y on X at values of Z:

$$\hat{Y} = (b_1 + b_3Z)X + (b_2Z + b_0) \tag{3.2}$$

There are four outcomes of rescaling by additive constants that will be shown algebraically for this equation using the approach of Cohen (1978).

1. In the case of linear regression with no higher order terms, that is, $b_3 = 0$ in equation 3.1, rescaling by additive constants has no effect on the value of the regression coefficients.
2. In regression equations containing at least one higher order term, rescaling by additive constants leads to changes in all regression coefficients except for the highest order term.
3. Simple slopes of simple regression equations are unaffected by additive transformations.
4. Under additive scale transformation the interpretation of the interaction as ordinal versus disordinal remains unchanged. The important conclusion from this exposition is that *our prescriptions for plotting and post hoc probing of interactions between continuous variables do not suffer from the problem of lack of invariance*, even though coefficients in the overall regression equation do.

Linear Regression with no Higher Order Terms

Transformation by additive constants has no effect on regression coefficients in equations containing only first order terms. To show this algebraically, we take the simple regression equation

$$\hat{Y} = b_1 X + b_2 Z + b_0 \quad (3.3)$$

and define two new variables $X' = X + c$ and $Z' = Z + f$, where c and f are additive constants. (Note that if c and f are the arithmetic means of X and Z, respectively, and X and Z represent centered variables, then X' and Z' represent the uncentered forms of these same variables). We rewrite the original centered variables as $X = X' - c$ and $Z = Z' - f$ and substitute these values into the simple regression equation, yielding:

$$\hat{Y} = b_1(X' - c) + b_2(Z' - f) + b_0 \quad \text{or}$$

$$\hat{Y} = b_1 X' + b_2 Z' + (b_0 - b_1 c - b_2 f) \quad (3.4)$$

Here, the coefficients b_1 and b_2 for the uncentered first order (X' and Z') terms are identical with those in equation 3.3 based on centered X and Z. Only the regression intercept $(b_0 - b_1 c - b_2 f)$ is changed from its original value.

Regression Equations with Higher Order Terms

It is quite a different matter if the regression equation contains an interaction or other higher order terms. Substituting the expressions $X = X' - c$ and $Z = Z' - f$ into equation 3.1 and collecting terms yields the following:

$$\hat{Y} = (b_1 - b_3 f)X' + (b_2 - b_3 c)Z' + b_3 X'Z' + (b_0 - b_1 c - b_2 f + b_3 cf) \quad (3.5)$$

Note that the original b_1 coefficient of equation 3.1 becomes $b_1' = (b_1 - b_3 f)$, the original b_2 coefficient of equation 3.1 becomes $b_2' = (b_2 - b_3 c)$, and the regression constant b_0 becomes $(b_0 - b_1 c - b_2 f + b_3 cf)$. Only the interaction coefficient does not change: b_3 thus retains its original value and interpretation.[1] The change in first order coefficients produced by linearly rescaling X and Z occurs when there is a nonzero interaction between these variables. The covariances between interaction term (XZ) and each component (X and Z) depend in part upon the means of the individual predictors. Rescaling changes the means, thus changes predictor covariances, resulting in changes in b_1 and b_2 for the predictors contained in the higher order function. This is true even if the individual predictors, X and Z, are uncorrelated with one another.[2] (The interested reader is referred to Appendix A, which provides algebraic expressions for the mean and variance of crossproduct terms XZ in terms of the means, variances, and covariances of its components X and Z. The covariance between a crossproduct term and its components is also explored.)

Simple Slopes of Simple Regression Equations

The simple slopes that are calculated from the interaction also remain constant under additive scale transformation. To see this algebraically, recall from equation 3.2 that $(b_1 + b_3Z)$ is the general form for the simple slopes of Y on X at values of Z from equation 3.1. Let us once again use the expressions $X = X' - c$ and $Z = Z' - f$, and substitute them into equation 3.2 for the regression of Y on X at levels of Z:

$$\hat{Y} = [b_1 + b_3(Z' - f)](X' - c) + [b_2(Z' - f) + b_0] \quad (3.6)$$

Expanding and collecting terms yields

$$\hat{Y} = (b_1 + b_3Z' - b_3f)X' + (-b_1c - b_3cZ' + b_3cf + b_2Z' - b_2f + b_0) \quad (3.7)$$

In order to compare the value of the simple regression coefficient $(b_1 + b_3Z)$ of equation 3.2 with the simple regression coefficient $(b_1 + b_3Z' - b_3f)$ in equation 3.7, we substitute the expression $Z' = Z + f$ into equation 3.7 with the result that

$$\hat{Y} = (b_1 + b_3Z)X' + (-b_1c + b_2Z - b_3cZ + b_0) \quad (3.8)$$

Note that the simple regression coefficient $(b_1 + b_3Z)$ for the regression of Y on X' at values of Z does not change from equation 3.2 to 3.8; rescaling predictors changes the regression constants but not the regression coefficients of the simple regression equations.

Ordinal Versus Disordinal Interactions

Once we have rescaled X and Z, the point at which the simple regression lines cross will move by the same factor as the additive constants c

and f for X and Z, respectively. To show this, we use the expression $b_2' = (b_2 - b_3 c)$ from equation 3.5 and recall that, for uncentered versus centered equation 3.1, $b_3' = b_3$. Substituting these expressions into equation 2.6, we find that the regression lines cross at the value

$$X'_{\text{cross}} = \frac{-b_2'}{b_3'} = \frac{-(b_2 - b_3 c)}{b_3} = \frac{-b_2}{b_3} + c \tag{3.9}$$

Thus transforming X by an additive constant moves the crossing points of the simple regression lines of Y on X by precisely the same constant. Hence the status of the interaction as ordinal versus disordinal is independent of the predictor scaling.

The algebraic relationships we have shown allow us to reach an important conclusion: Any additive transformation of the original variables has no effect on the overall interaction or on any aspect of the interaction we might choose to examine.

Numerical Example— Centered Versus Uncentered Data

Our examination of Tables 2.1, 2.2, and 2.3 in the previous chapter was focused solely on those portions of the tables that report the results of analyses using centered variables X and Z. Also contained in these tables are the results of parallel analyses in which the variables have been transformed to their uncentered forms as follows: $X' = X + 5$ and $Z' = Z + 10$. This example permits the direct comparison of the regression analysis, plots, and post hoc probing based on centered versus uncentered data. It also introduces some of the desirable properties of centered solutions.

Correlations

Note in Table 2.1a that the correlations between the centered terms X and XZ and between Z and XZ are low, .10 and .04, respectively. However, in the uncentered case, large correlations are introduced between X' and $X'Z'$ and between Z' and $X'Z'$. For example, the correlation between uncentered Z' and $X'Z'$ is .86, instead of .04 for Z with XZ. This example illustrates how considerable multicollinearity can be introduced into a regression equation with an interaction when the variables are not centered (Marquardt, 1980). Very high levels of multicollinearity can lead to technical problems in estimating regression coefficients. Centering variables will often help minimize these problems (Neter, Wasserman, & Kutner, 1989).

Regression Equations with no Higher Order Terms

The regression equations including first order terms only (no interaction) are given in Tables 2.1c(i) and 2.1d(i) for centered versus uncentered data. Note that the regression coefficients $b_1 = 1.67$ and $b_2 = 3.59$ are identical for the centered and uncentered equations. Only the regression constant reflects the change in scaling.

Regression Equations Containing an Interaction

Comparing Tables 2.1c(ii) and 2.1d(ii), we note that the b_3 coefficient for the interaction term is identical in the uncentered and centered equations, as are the tests of significance. However, in the uncentered equation, both of the coefficients for X' and Z' are negative and are significant. In sharp contrast, in the centered equation, both of these coefficients are positive, with the coefficient for Z being significant. Such dramatically different results with simple additive rescaling of the data highlight the difficulties of regression with interactions. As has already been shown algebraically, the equivalence of the centered and uncentered analyses is clarified in the simple slope analysis.

Simple Slopes

Equation 3.2 expresses the regression of Y on X at particular values of Z. Using uncentered data, we compute the simple slope equations at the values $Z'_{H} = \bar{Z}' + 1$ standard deviation, $Z'_{M} = \bar{Z}'$, and $Z'_{L} = \bar{Z}' - 1$ standard deviation. To calculate the simple slope equations in the uncentered case, we use the uncentered regression equation in Table 2.1d(ii) containing the interaction:

$$\hat{Y} = -24.68X' + -9.33Z' + 2.58X'Z' + 90.15$$

This equation of Table 2.1d(ii) is reexpressed in Table 2.2b in the form of equation 3.2; that is,

$$\hat{Y} = (-24.68 + 2.58Z')X' + (-9.33Z' + 90.15)$$

Substituting the value for uncentered $Z'_{H} = 12.20$, we have $\hat{Y} = [-24.68 + 2.58(12.20)]X' + [(-9.33)(12.20) + 90.15] = 6.82X' - 23.67$. For centered $Z_{H} = 2.20$, recall that $\hat{Y} = [1.14 + 2.58(2.20)]X +$

$[3.58(2.20) + 2.54] = 6.82X + 10.41$. Comparison of the simple slopes for Z_H, Z_M, and Z_L for the centered versus uncentered solutions in Table 2.2a versus 2.2b shows an important result: The corresponding simple slopes are identical. That is, the centered and uncentered simple slope equations for a value of Z of the same relative standing across equations (e.g., Z_H one standard deviation above the mean) have the same slopes. Hence the relationship between Y and X is unambiguously portrayed in the simple slope equations regardless of whether these simple slope equations are generated from the centered or uncentered equations. A comparison of Figure 2.1a versus 2.1b verifies the equivalence of the simple slopes from the interaction generated from centered versus uncentered data.

Standard Errors of Simple Slopes and t-Tests

The standard errors of simple slopes and hence the t-tests are also invariant under additive transformation. The variance-covariance matrix of the uncentered regression coefficients, $\mathbf{S}_{b'}$, is given in Table 2.3b. This matrix was obtained from the regression program of the SPSS-X package applied to the uncentered data set. The square root of equation 2.4 for the variance of the simple slope, that is, $[s_b = (s_{11} + 2Zs_{13} + Z^2 s_{33})^{1/2}]$, applies to both centered and uncentered data. For example, s'_H, the estimate of the standard error of the simple slope of Y on X' at $Z'_H = 12.20$ is given as $[43.88 + (2)(12.2)(-4.07) + 12.2^2(0.40)]^{1/2} = 1.98$, where $s'_{11} = 43.88$, $s'_{13} = -4.07$, and $s'_{33} = 0.40$ from the matrix $\mathbf{S}_{b'}$. For centered $Z_H = 2.20$, b_H was found to be $[2.35 + 2(2.2)(-0.08) + 2.2^2(0.40)]^{1/2} = 1.98$. The simple slopes, standard errors, and t-tests based on centered versus uncentered data are identical, as is shown in Table 2.3.

Ordinal Versus Disordinal Interactions

In order to examine the effect of centering on the ordinal versus disordinal status of the interaction, we must determine the effect of the transformation on the crossing point of the simple regression lines. Applying equation 2.6, that is, $X'_{\text{cross}} = -b'_2/b'_3$, for the value of X' at which all regressions of Y on X' will cross, we find $X'_{\text{cross}} = -(-9.33)/2.58 = 3.61$. With $\overline{X} = 5.0$ and $s'_X = 0.95$, we see that the regressions of Y on X' cross at $(3.61 - 5)/0.95 = -1.47$ standard deviations below the mean. We have already seen in Chapter 2 that for the centered solution the crossing point also falls precisely 1.47 standard deviations below the mean of X (see p. 24).

Should the Criterion Y be Centered?

Throughout Chapters 2 and 3 we have left the criterion Y uncentered, even in analyses with centered predictors. Changing the scaling of the criterion by additive constants has no effect on regression coefficients in equations containing interactions. By leaving the criterion in its original (usually uncentered) form, predicted scores conveniently are in the original scale of the criterion. There is typically no reason to center the criterion Y when centering predictors.

Multicollinearity: Essential Versus Nonessential Ill-Conditioning

If the first order variables X and Z are not centered, then product terms of the form XZ and power polynomial terms of the form X^2 are highly correlated with the variables of which they are comprised (the Pearson product moment correlation between X and X^2 can approach 1.0). When employed in regression analyses with lower order terms, the highest order term produces large standard errors for the regression coefficients of the lower order terms, though the standard error of the highest order term is unaffected. Cohen (1978) and Pedhazur (1982) acknowledge the computational problems that may arise from this multicollinearity.

The literature on regression with higher order terms contains many admonitions about the problems of multicollinearity. However, these problems are not the usual problems of multicollinearity in regression analysis in which two supposedly different predictors are very highly correlated. The multicollinearity in the context of regression with higher order terms is due to scaling, and can be greatly lessened by centering variables. The special cases we consider in this book (e.g., the relationship between X and X^2, or between X and Z and their product XZ) follow a general result. Uncentered X' and X'^2 will be highly correlated. But if instead we use centered predictor X and it is normally distributed, then the covariance between centered predictor X and X^2 is zero. Even if X is not normally distributed, the correlation between X and X^2 will be much lower than the correlation between X' and X'^2. Uncentered X' and Z' will both be highly correlated with their crossproduct $X'Z'$. But if X and Z are bivariate normal, then the covariance between each centered variable X and Z and the product term XZ is zero. When X and Z are centered, the only remaining correlation between first order and product terms or between first order and second order terms is that due to nonnormality of the variables. (We again recommend Appendix A for the mathematical basis of these statements.)

Marquardt (1980) refers to the problems of multicollinearity produced by noncentered variables as *nonessential ill-conditioning*, whereas those that exist because of actual relationships between variables in the population (e.g., between the age of a child and his/her developmental stage) are referred to as *essential ill-conditioning*. Nonessential ill-conditioning is eliminated by centering the predictors.

We recommend centering for computational reasons. Marquardt (1980), Smith and Sasaki (1979), and Tate (1984) provide clear discussions of approaches to reducing multicollinearity in interactive regression models (see also Lance, 1988).

Interpreting the Regression Coefficients

The Interaction Term XZ

At the beginning of Chapter 2, we pointed out that an interaction between continuous predictors indicates that the regression of the criterion on each of the predictors varies as a function of the value of the other predictor. As seen in the simple slope expression contained in equation 3.2, the value of the b_3 regression coefficient for the product term indicates the amount of change in the slope of the regression of Y on X that results from a one-unit change in Z (see also Cleary & Kessler, 1982; Finney, Mitchell, Cronkite, & Moos, 1984; Judd & McClelland, 1989). In terms of the simple regression equations[3] in Table 2.2, the slope of the regression of Y on X at levels of Z increases by 2.58 units for every one unit increase in Z. Note that this change is monotonic and completely uniform across the range of Z. The simple product term XZ represents an interaction that always appears like a fan when plotted as a series of simple regression equations such as those in Figure 2.1. Interactions representing curvilinear and nonuniform effects can also be built into regression equations; however, we defer the discussion of these more complex interactions until Chapter 5.

We have already seen that the product term XZ in equation 3.1 is unaffected by the scale of measurement. Hence, its interpretation holds across additive scale transformations. Note, however, that this constancy pertains only to the *unstandardized* b_3 coefficient; the standardized coefficient (beta) for the interaction is affected by additive transformation as will be shown later in this chapter.

The First Order Terms X and Z

Centered Versus Uncentered Variables

In a regression equation containing an interaction the regression coefficients for the first order terms (i.e., b_1 and b_2 in equation 3.1) are examples of what have been labeled *conditional effects* (e.g., Cleary & Kessler, 1982; Darlington, 1990). Conditional effects describe the effect of one predictor on the criterion variable under the condition in which the other predictor equals a specified value. For a conditional effect to be useful, the point on the other predictor at which it is evaluated must be meaningful.

In equation 3.1, the b_1 coefficient represents the regression of Y on X at $Z = 0$; the b_2 coefficient represents the regression of Y on Z at $X = 0$. These coefficients will not always be meaningful for uncentered predictors. For example, if strength of athletes were predicted from their height (X) and weight (Z), the regression coefficient predicting strength from height (b_1) would represent the regression of strength on height for athletes weighing zero pounds. Often in social science research X and Z are measured on interval scales in which the value zero has no meaning. If some behavior were predicted from a measure of motivation (X) and a 7-point attitude scale (Z) ranging from 1 to 7, the regression coefficient for Y on X would be the slope of Y on X at the value $Z = 0$, a value not even defined on the scale! However, when predictors are centered, then the value of 0 is the mean of each predictor. Hence, if Z is centered, then the b_1 coefficient for X represents the regression of Y on X at the mean of the Z variable. Centering produces a value of zero on a continuous scale that is typically meaningful.

The relationship between the b_1 coefficient in a centered overall regression equation and the simple slope analysis may now be clarified: The b_1 coefficient from the overall centered regression equation is equal to the simple slope for the regression of Y on X at the mean of Z (Z_M). In Table 2.3, the value $b_M = 1.14$ is simple slope of Y on X at Z_M (equivalently, from the uncentered equation, $b'_M = 1.14$). We have already encountered the concept of the regression of Y on X at the mean of Z in the simple regression equation analysis portrayed in Tables 2.2 and 2.3. In Table 2.2, the simple slope of the regression of Y on X at Z_M or equivalently of Y on X' at Z'_M is 1.14. From Table 2.1c(ii), the regression coefficient b_1 is equal to 1.14 in the overall centered equation.

There is even more convergence between the b_1 coefficient in the centered analysis and the simple regression analyses summarized in Table 2.3. From the *centered* matrix $\mathbf{S}_b$ the variance of the b_1 coefficient in the overall centered regression equation is $s_{11} = 2.35$; the standard error is thus $2.35^{1/2} = 1.53$, and $t = 1.14/1.53 = 0.74$. These are precisely the same as the standard error and t value for the simple slope b_M or b'_M in Table 2.3. Thus there is a clear interpretation of each of the b coefficients in the centered regression equation.

In the uncentered equation the b'_1 coefficient does retain its interpretation as the regression of Y on X' at $Z' = 0$. However, the value of zero is no longer at the center of the data, and, in fact, may not even exist on the scale of variable Z. Only in the special case in which uncentered variable Z' has a meaningful zero point is the regression coefficient for X' meaningful. Only when the uncentered X' variable has a meaningful 0 point is the b'_2 coefficient meaningful. Because the centered overall regression analysis provides regression coefficients for first order terms that may be informative, we recommend that the centered analysis be employed, echoing the recommendations of Finney et al. (1984) and Marquardt (1980). This also conforms to the familiar model used in ANOVA: Each main effect is estimated when the value of all other factors are equal to their respective means.

Interpretation in the Presence of an Interaction.

The b_1 and b_2 coefficients in equation 3.1 do not represent "main effects" as this term is usually used. Main effects are most typically defined as the *constant* effect of one variable across all values of another variable (Cramer & Appelbaum, 1980). Less commonly they are defined as *average* effect of one variable across the range of other variables (Finney et al., 1984). Darlington (1990) defines average effect as the average of the simple slopes computed across all cases. In other words, if one were to substitute observed predictor scores for each case into the simple slope expression ($b_1 + b_3Z$) for the regression of Y on X, calculate for each case a value of the simple slope, and then average these across all cases, this would be considered the average effect of Y on X. For a centered regression equation, the average effect of Y on X is b_1. Put another way, if one calculated a simple slope of Y on X at every value of Z, weighted each such slope by the number of cases with that value of Z, and took the weighted average of the simple slopes, the result would be the average simple slope, equal to b_1 in the centered regression equation.

The b_1 and b_2 coefficients never represent constant effects of the predictors in the presence of an interaction. The b_1 and b_2 coefficients from centered equations always represent the effects of the predictors at the mean of the other predictors. In the centered equation, they may also be considered as the weighted average effect of each predictor coefficient across all observed values of the other predictor. The interpretation of b_1 or b_2 as conditional effects of predictors at the mean of other predictors may well be useful in clarifying relationships under investigation. Hence we agree with the position of Finney et al. (1984) that these effects should not be disregarded simply because they are not constant effects. We also echo the admonition of Cleary and Kessler (1982) that the interpretation of these effects warrants careful consideration of the scale characteristics of the variables they represent.

Given that the b_1 coefficient for centered Z always represents the regression of Y on X at the mean of Z, the range of Z should be explored to determine over what range of Z the relationship represented by the b_1 coefficient holds. Consider the centered regression equation $\hat{Y} = 3X + 2Z + 0.5XZ + 5$, rearranged as $\hat{Y} = (3 + 0.5Z)X + (2Z + 5)$, with Z centered and $s_Z = 1.5$. The regression coefficient for Y on X equals 3 when $Z = 0$ (i.e., $b_1 = 3$). As Z increases above zero, the regression of Y on X becomes increasingly positive; the value of the simple slope $(3 + 0.5Z)$ increases as Z increases. When Z decreases to $Z = -1.5$ (one standard deviation below its mean), the simple slope for Y on X is still positive $[\text{i.e.}, 3 + 0.5(-1.5) = 2.25]$, as it is when $Z = -3.0$, two standard deviations below the mean of Z. Thus the conclusion that, on average, the regression of Y on X is positive across the range of Z is quite accurate. In contrast, the b_1 coefficient in Table 2.1c(ii) represents neither the regression of Y on X one standard deviation above the mean of Z nor that regression at one standard deviation below the mean of Z.

We recommend that in characterizing the first order effects of a regression analysis containing an interaction, consideration be given to the range of each variable over which the first order effect of the other variable holds true. This eliminates the need to follow some rigid rule about considering versus not considering the conditional effects of first order variables and provides an accurate picture of the outcome.

A Geometric Interpretation

A geometric representation of the b_1 and b_2 conditional effects and the b_3 interaction (Cleary & Kessler, 1982) provides insight into their meaning. The b_2 coefficient for Z indicates how the predicted score $\hat{Y}$ changes as a function of Z at $X = 0$. In Figure 2.1a, the value $X = 0.0$ is shown in the center of the X axis. Reading up from the X axis at $X = 0.0$ to a simple regression line at Z_L and then across from the simple regression line to the Y axis yields a predicted score $\hat{Y}$ at $X = 0.0$ and $Z = Z_L$. The predicted scores at $X = 0.0$ are -5.33, 2.54, and 10.41 for Z_L, Z_M, and

Z_H, respectively; these values are the regression constants (Y intercepts) in the simple regression equations. Note that as Z increases, $\hat{Y}$ increases; in the centered regression equation $b_2 = 3.58$ is positive.

Now consider Figure 3.1, which is an expanded illustration of Figure 2.1b for the uncentered data. In Figure 3.1, we have extended the X axis downward to 0.0 and have extended the simple regression lines to intercept the Y axis. The b_2' coefficient in the uncentered regression equation still is interpreted at the regression of Y on Z' at $X' = 0.0$. Reading up from $X' = 0.0$ to the three simple regression equations yields values of $\hat{Y}$ at values of Z'. The predicted scores are 17.38, −3.15, and −23.67 for Z_L', Z_M', and Z_H', respectively. As Z' increases, $\hat{Y}$ decreases, and $b_2' = -9.33$ is negative in the uncentered regression equation. Rescaling by additive constants produces shifts in the origin of the regression plane; in the example the change in sign of b_2 from centered to uncentered equation is occasioned by the fact that the simple regression lines do not cross above zero in the centered case but do in the uncentered case. Finally, this shift in origin explains the difference in Y intercepts of the simple slope equations between the centered and uncentered solutions.

What of b_3? The b_3 coefficient represents the angle between the regression lines of Y on X at values of Z or Y on Z at values of X. These angles do not change with shifts in the origins of the axes produced by rescaling.

Standardized Solutions with Multiplicative Terms

To this point we have come to expect that changing the scale of predictor variables by additive constants, as in centering, will have no effect on the unstandardized regression coefficient for the interaction term. But what of the standardized regression coefficients (betas) for the interaction term associated with centered versus uncentered scores? Table 3.1 compares the unstandardized regression coefficients for the uncentered and centered solutions of the numerical examples (presented as Cases 1a and 2a in Table 3.1). Standardized regression coefficients associated with each of these analyses are also presented (Cases 1b and 2b of Table 3.1, respectively). The comparison shows one reassuring and two disconcerting findings.

1. The t-tests of the standardized regression coefficients for the interaction term in the centered versus noncentered analyses are identical ($t = 4.087$).
2. The standardized regression coefficients for the interaction term in the centered versus uncentered analyses differ substantially (1.61 versus 0.19).

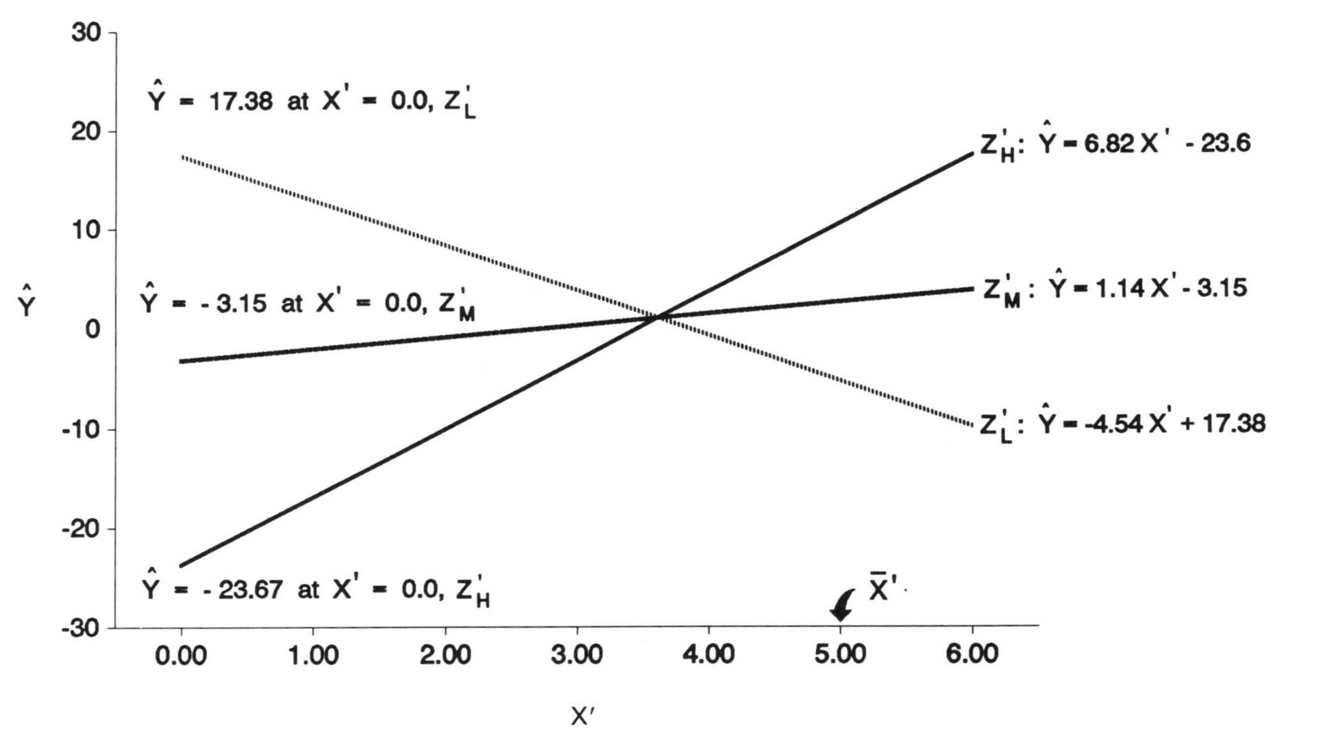

Figure 3.1. Interaction Plotted from Uncentered Regression Equation: $\hat{Y} = -24.68X' - 9.33Z' + 2.58X'Z' + 90.15$

3. The simple slopes generated from the two standardized solutions are also substantially different. To illustrate, let us compute the simple slope for the regression of Y on X, that is, $(b_1 + b_3Z)$, at $Z_H = Z$ one standard deviation above the mean (i.e., when $Z = 1$ for the standardized case). For solution 1b of Table 3.1,

$$(b_1 + b_3Z) = [-0.83210 + 1.61337(1)] = 0.78127$$

whereas for solution 2b of Table 3.1,

$$(b_1 + b_3Z) = [0.03832 + 0.19218(1)] = 0.23050$$

In this section, following the development of Friedrich (1982; see also Jaccard, Turrisi, & Wan, 1990), we explore why standardized regression coefficients (betas) for interaction terms and simple slopes do not display

Table 3.1
Raw and Standardized Solutions Based on Centered versus Uncentered Data

	Analysis	b_1 *(for X)*	b_2 *(for Z)*	b_3 *(for XZ)*	b_0	*t-test for* b_3	*Simple slope at* Z_H $(b_1 + b_3Z_H)$
1a.	Raw uncentered Y, X, and Z	−24.67759	−9.32984	2.58141	90.15337	4.087	
1b.	Standardized solution associated with 1a (i.e., raw uncentered Y, X and Z)	−0.83210	−0.73258	1.61337	—	4.087	0.78127[b]
2a.	Raw centered X and Z	1.13648	3.57720	2.58141	2.53743[a]	4.087	
2b.	Standardized solution associated with 2a (i.e., raw centered X and Z)	0.03832	0.28088	0.19218	—	4.087	0.23050[b]
3a.	Standardized z_Y, z_X, z_Z, and (z_Xz_Z) as predictors, raw analysis	0.03832	0.28088	0.19150	−0.07930	4.087	0.22982
3b.	Standardized solution associated with 3a	0.03832	0.28088	0.19218	—	4.087	0.23050[b]

[a] Criterion Y is uncentered, in order to provide predicted scores in the original scale of the criterion.
[b] These outcomes are the result of inappropriate factorization and should not be employed; see text.

the invariance properties we have come to expect. The strong implication of this lack of invariance is that neither the traditional standardized centered nor the standardized uncentered solution (solutions 1b and 2b in Table 3.1) should be used when interactions are present.

Appropriate Standardized Solution with Interaction Terms

The computation of standardized regression coefficients (betas) in a typical computer analysis begins with the standardization of each of the predictors. (Whereas this is correct for regression equations that do not contain interactions, it is not so for regression equations containing interactions, as we will see.) For the raw uncentered analyses, these predictors are X', Z', and $X'Z'$, and the corresponding z-scores are $z_{X'}$, $z_{Z'}$, and $z_{X'Z'}$. The final term $z_{X'Z'}$ is the z-score computed from the raw product term $X'Z'$; that is, the raw term is formed, and then it is standardized. It is *not* in general equal to the product of z-scores $(z_{X'}z_{Z'})$. The same is true for the centered raw score analysis. The input variables are centered X, centered Z, and their product XZ. For the standardized solution, the predictor for the interaction is z_{XZ}, the z-score computed from the product of centered raw scores. Again, it is not necessarily equal to the product of z-scores $(z_X z_Z)$.

To compute the simple slopes, we must be able to factor the product terms. Recall that to find the expression for the simple slope of the regression of Y or X at levels of Z in the unstandardized case, we factor the XZ term:

$$\hat{Y} = b_1X + b_3XZ + b_2Z + b_0 = (b_1 + b_3Z)X + b_2Z + b_0$$

The same is true for the standardized analysis

$$\hat{z}_Y = b_1^* z_X + b_2^* z_Z + b_3^* z_X z_Z + b_0^*$$

$$\hat{z}_Y = (b_1^* + b_3^* z_Z) z_X + b_2^* z_Z + b_0^*$$

where the b_i^* are standardized regression coefficients and $\hat{z}_Y$ is the predicted standard score corresponding to Y. In order for the required factoring to be performed, the predictor variable for the interaction term must be the crossproduct of z-scores, but this is not the case in either standardized solution 1b or 2b in Table 3.1 because the crossproduct terms have been standardized.

Friedrich (1982) suggested a straightforward procedure that solves this problem. One first calculates z_X and z_Z and then forms their crossproduct $z_X z_Z$. These values are used as the predictors in a regression analysis, with z_Y as the criterion. The *unstandardized* solution from that analysis is the appropriate "standardized" solution for use with multiplicative terms. This solution is given in Table 3.1(3a). Note that in this analysis the regression intercept b_0 will typically be nonzero, though in traditional standardized regression analyses this coefficient is always zero.

To understand this difference in intercepts, recall that in general $b_0 = \bar{Y} - b_1\bar{X}_1 - b_2\bar{X}_2 \cdots - b_K\bar{X}_K$. In the traditional additive standardized solution all variables have mean zero. In the procedure suggested by Friedrich (1982) presented in Table 3.1(3a), the crossproduct term $z_X z_Z$ does not have a mean of zero. If two centered variables are crossmultiplied, the mean of the resulting crossproduct term equals the covariance between the variables (see Appendix A). In the special case of two standardized variables, the mean of the crossproduct term equals their correlation. Thus the mean of the crossproduct terms $z_X z_Z$ will be zero only when X and Z are uncorrelated. Similarily, the variance of the crossproduct term XZ is a function of both the variances of X and Z and the covariance between X and Z. (In the case of bivariate normal X and Z, $\sigma^2_{XZ} = \sigma^2_X\sigma^2_Z + \text{Cov}^2_{XZ}$; see Appendix A.) The variance of the crossproduct of two z-scores will equal 1 only if z_X and z_Z are uncorrelated. In all, then, the crossproduct of two standardized variables z_X and z_Z is itself standardized only if X and Z are uncorrelated. With the mean of the product term $z_X z_Z$ equal to r_{XZ}, the correlation between X and Z, the value of the intercept in solution 3a is $-b_3^* r_{XZ}$.

Note that the *unstandardized solution* must be used with Friedrich's procedure. The standardized solutions (betas) associated with this procedure (Table 3.1(3b)) present the same problems as those given in 1b and 2b: The interaction term is not a crossproduct of z-scores, rather it is the z-score calculated from the crossproduct of z-scores. For the same reason that solutions 1b and 2b are inappropriate as standardized solutions, solution 3b will also be inappropriate.

Simple Slope Analysis from the Standardized Solution

The "standardized" solution from the Friedrich procedure is given in Table 3.1(3a) as follows:

$$\hat{z}_Y = 0.03832\, z_X + 0.28088\, z_Z + 0.19150\, z_X z_Z - 0.07930$$

Following the procedures developed in Chapter 2, we can treat this "standardized" solution just like any other regression equation and calculate the simple slopes of z_X at high (+1), moderate (0), and low (−1) levels of z_Z. The standard errors and *t*-tests can then be performed by substituting the appropriate values in the formulas presented in Chapter 2 (equations 2.4 and following text). These results are presented in Table 3.2; Table 2.3a presents the results of the same analysis performed on the raw centered data.

Relationship Between Raw and "Standardized" Solution

There are simple algebraic relationships between the centered raw score analysis with centered *X*, *Z*, and their crossproduct *XZ* as predictors and the "standardized" analysis using the Friedrich procedure with z_X, z_Z, and $z_X z_Z$ as predictors. These relationships are presented below and in Table 3.2 for the values related to tests of simple slopes.

1. The regression coefficients are related as follows:

$$b_i^* = b_i \frac{s_i}{s_Y}$$

where b_i^* is the standardized regression coefficient associated with predictor *i*, b_i is the unstandardized regression coefficient associated with predictor *i*, s_i is the standard deviation of predictor *i*, and s_Y is the standard deviation of the criterion *Y*. For example, from Table 3.1(2a), b_3 = 2.58141; from Table 3.1(3a) $b_3^* = 0.19150$, $s_Y = 28.01881$, and s_{XZ} = 2.08592; or $b_3^* = 2.58141\ (2.08592/28.01881)$. This relationship holds for all the regression coefficients in the equation. It is the usual relationship that is obtained for any linear regression analysis involving only first order terms (see, e.g., Cohen & Cohen, 1983).

2. The regression constants (intercepts) in the two analyses have the following relationship:

$$b_0^* = \frac{b_0 - \overline{Y}}{s_Y}$$

where b_0^* and b_0 are the standardized and unstandardized constants, respectively, and $\overline{Y}$ is the mean of the criterion scores. If *Y* has been centered, then this reduces to $b_0^* = (b_0)/s_Y$.

3. The variance-covariance matricies of the regression coefficients in the two solutions are related as follows:

$$\text{standardized element}_{ij} = \text{unstandardized element}_{ij} \left(\frac{s_i s_j}{s_Y^2} \right)$$

where i and j refer to any two predictors, s_i and s_j are their respective standard deviations, and s_Y^2 is the variance of the criterion. This relationship may be verified numerically by comparing the values for the standardized solution in Table 3.2(a) to those for the unstandardized solution in Table 2.3(a).

Table 3.2
Simple Slope Analysis Based on Predictors z_X, z_Z, and $z_X z_Z$ (for Comparison with Raw-Centered Simple Slope Analysis in Table 2.3a)

a. Covariance Matrix of Regression Coefficients

	b_1	b_2	b_3
b_1	.00267	−.00110	−.00021
b_2	−.00110	.00265	−.00001
b_3	−.00021	−.00001	.00220

Relationship to covariance matrix in Table 2.3a

$$\text{Standardized element}_{ij} = \text{Raw element}_{ij} \left[\frac{s_i s_j}{s_Y^2} \right]$$

where s_i and s_j are the standard deviations of the raw predictors

b. Simple slopes

$b_L^* = -.15332$
$b_M^* = .03832$
$b_H^* = .22982$

Relation to simple slopes of raw-centered analysis in Table 2.3a

$$\text{Standardized simple slope of } Y \text{ on } X = \text{Raw simple slope of } Y \text{ on } X \left[\frac{s_X}{s_Y} \right]$$

c. Standard Errors of Simple Slopes

$s_L^* = .07267$
$s_M^* = .05167$
$s_H^* = .06678$

Relation to standard errors in raw-centered analysis in Table 2.3a

$$\text{Standardized simple standard error} = \text{Raw simple standard error} \left[\frac{s_X}{s_Y} \right]$$

d. t-tests

$t_L^* = -2.11$
$t_M^* = 0.74$
$t_H^* = 3.45$

Relationship to t-tests in raw-centered analysis in Table 2.3a

t-tests are identical in raw-centered and standardized analyses

NOTE: $s_X = 0.94476$; $s_Z = 2.20004$; $s_{XZ} = 2.08592$; $s_Y = 28.01881$

4. From (3), it follows that the standard errors of the regression coefficients are related as follows:

$$s_b^* = s_b \frac{s_i}{s_Y}$$

where s_b^* and s_b are the standard errors of the standardized and raw regression coefficient for predictor i, respectively, and s_i is the standard deviation of predictor i.

5. From (3), it also follows that the matrices of correlations among the predictors are identical for the unstandardized and standardized solutions.

6. From (1) and (4), it follows that the t-test values and p values for tests of the regression coefficients in the two analyses are identical.

7. Finally, Table 3.2 compares the simple slopes, the standard errors of the simple slopes, and the t-tests for simple slopes for the unstandardized and standardized solutions. As can be seen, both the simple slope and standard error of the simple slope are related to their unstandardized counterparts by identical algebraic expressions. The t-tests for the simple slopes for the standardized and unstandardized centered cases are identical.

In summary, using the Friedrich (1982) approach to standardization preserves the usual relationships between raw and standardized solutions found for regression equations that involve only linear terms. These relationships are also preserved for the simple slope analyses. Thus, of the four possible standardized solutions presented in Table 3.1, only the Friedrich approach (3a) is algebraically appropriate and bears a straightforward algebraic relationship to the unstandardized centered analysis in (2a). Remember in this approach that the predictors are all z-scores or their products to begin with; they should not be further standardized. The use of this approach avoids potential computational difficulties and ambiguities of interpretation.

Summary

This chapter has addressed the issue of scale invariance when there are interactions between continuous variables in MR. The discrepancies in the regression coefficients obtained from centered and uncentered analyses vanish when our prescriptions for probing the interaction are followed. Centered and uncentered analyses lead to identical slopes of the simple regression equations and identical tests of the highest order interaction. The interpretation of first order terms in regression equations containing interactions is considered; such first order terms represent *conditional* rather than constant effects of single predictors. Our consideration of the interpretation of these coefficients clarifies an advantage of centering variables before analysis, as did our consideration of multicollinearity between lower and higher order terms. Problems associated with standardized solutions of regression equations containing interactions are discussed and an appropriate standardized solution is presented.

Notes

1. The constancy of the unstandardized b_3 coefficient across the centered and uncentered analyses does not hold for b_3 in the *standardized* solutions based on centered versus uncentered data (see the final section of this chapter).

2. These comments concerning the interaction pertain to the XZ term in equation 3.1. In equations with higher order terms such as X^2 and X^2Z, considered in Chapter 5, only the regression coefficient for the highest order term is invariant under linear scale transformation.

3. The simple slopes for the regression of Y on Z at values of X from equation 3.1 is $(b_2 + b_3X) = (3.58 + 2.58X)$ in the numerical example. For every one unit increase in X, there will be 2.58 units of increase in Z. Hence there is a symmetry of the regressions of Y on X at Z and Y on Z at X.

4 Testing and Probing Three-Way Interactions

Chapters 1–3 have focused exclusively on interactions between two predictor variables. The present chapter shows how the prescriptions for testing, interpreting, and probing XZ interactions developed in previous chapters generalize immediately to the three variable case. We limit our treatment here to interactions involving only linear terms (XZW); the discussion of interactions involving higher order, curvilinear components is deferred until Chapter 5.

Specifying, Testing, and Interpreting Three-Way Interactions

The usual requirement for developing a regression equation that includes a three-way interaction is that all first order and second order terms must be included in the equation.[1] As before, each of the predictor variables should be centered to maximize interpretability and to minimize problems of multicollinearity. The predictor for the three-way interaction is formed by multiplying together the three predictors. These considerations result in the following regression equation:

$$\hat{Y} = b_1X + b_2Z + b_3W + b_4XZ + b_5XW + b_6ZW + b_7XZW + b_0 \qquad (4.1)$$

In this equation, the test of the b_7 coefficient indicates whether the three-way interaction is significant. The two-way interactions (e.g., XZ) now represent conditional interaction effects, evaluated when the third variable (e.g, W) equals 0. They are affected by the scale of the predictor just as are first order terms X and Z in the presence of the XZ interaction. With centered predictor variables, the two-way interactions are interpreted as conditional interaction effects at the mean of the variable not involved in the interaction (e.g., the conditional XZ interaction at the mean of W). First order effects (e.g., X) may also be interpreted as conditional effects (e.g., when W and $Z = 0$; see pp. 37–40). If the XZW interaction in equation 4.1 is significant, then this interaction should be probed to assist in its interpretation. If the highest order interaction in the regression equation is not significant, readers may wish to use the stepdown procedures presented in Chapter 6.

Probing Three-Way Interactions

Simple Regression Equation

We begin by developing the simple regression equation, rearranging equation 4.1 to show the regression of Y on X:

$$\hat{Y} = (b_1 + b_4 Z + b_5 W + b_7 ZW)X + (b_2 Z + b_3 W + b_6 ZW + b_0) \tag{4.2}$$

Equation 4.2 shows that the regression of Y on X given in the expression $(b_1 + b_4 Z + b_5 W + b_7 ZW)$ depends upon both the value of W and the value of Z at which the Y on X relationship is considered. Equation 4.2 is now in the form of a *simple regression equation*, a direct generalization from the simple regression equation $Y = (b_1 + b_3 Z)X + b_2 Z + b_0$ for the two predictor case. The expression $(b_1 + b_4 Z + b_5 W + b_7 ZW)$ is the simple slope of the new simple regression equation.

Numerical Example

A simulation involving three multivariate normal predictors X, Z, and W is used to illustrate the probing of the three-way interaction. Table 4.1 provides the regression analysis and computation of simple slopes. Table 4.3a provides the means and standard deviations of predictors and criteria; note that the first order predictors X, Z, and W are centered but the criterion and the crossproduct terms are not. Table 4.1a gives the overall regression analysis; the three predictor XZW interaction term is significant. In Table 4.1b, the overall regression equation is rearranged according to equation 4.2 to show the regression of Y on X at levels of Z and W. In Table 4.1c, four separate simple regression equations are generated, one at each of the four combinations of Z and W one standard deviation above and below their means, that is, at combinations of Z_L and Z_H with W_L and W_H. For example, for the simple regression equation at Z_H and W_H, with centered Z and W, and $s_Z = 3.096$, $s_W = 1.045$, the values $Z_H = 3.096$ and $W_H = 1.045$ were substituted into the equation in Table 4.1b. This substitution is shown below:

$$\begin{aligned}\hat{Y} = \big[&-0.7068 + (0.5234)(3.096) + (1.0007)(1.045)\\&+ (0.7917)(3.096)(1.045)\big]X + \big[(2.8761)(3.096)\\&+ (14.2831)(1.045) + (-1.7062)(3.096)(1.045)\\&+ 4.5710\big]\end{aligned}$$

$$\hat{Y} = 4.521X + 22.881$$

The result is the simple regression equation for Y on X at Z_H, W_H, given in Table 4.1c(i). The results of the remaining three substitutions are given in Table 4.1c as well.

Table 4.1
Three Predictor Regression Analysis

a. Overall Regression Equation

$$\begin{aligned}\hat{Y} = &-0.7068X + 2.8761Z + 14.2831W^{*} + 0.5234XZ^{**}\\&+ 1.0007XW - 1.7062ZW + 0.7917XZW^{***} + 4.5710\end{aligned}$$

b. Regression Equation Rewritten to Show Simple Regression Equation of Y on X at Values of Z and W

$$\begin{aligned}\hat{Y} = &(-0.7068 + 0.5234Z + 1.0007W + 0.7917ZW)X\\&+ (2.8761Z + 14.2831W + -1.7062ZW + 4.5710)\end{aligned}$$

c. Simple Regression Equations at Values One Standard Deviation Above and Below the Means of Z and W, where $s_Z = 3.096$, $s_W = 1.045$

(i) At Z_H and W_H: $\hat{Y} = 4.521X + 22.881$

(ii) At Z_L and W_H: $\hat{Y} = -3.843X + 16.112$

(iii) At Z_H and W_L: $\hat{Y} = -2.693X + 4.070$

(iv) At Z_L and W_L: $\hat{Y} = -0.811X - 24.779$

$^{***}p < .001$; $^{**}p < .01$; $^{*}p < .05$.

Graphing the Three-Way Interaction

Casting of the three predictor interaction into a series of simple regression equations permits the plotting of the interaction. Figure 4.1 illustrates the three predictor XWZ interaction. The plots shown should appear highly familiar to readers who have worked with plots of three-factor interactions in the ANOVA context. The two graphs of Figure 4.1 are generalizations of Figure 2.1, in which the XZ interaction was plotted to show the regression of Y on X at levels of the second variable Z. Each graph in Figure 4.1 also shows the regression of Y on X at levels of the second variable Z. The third variable W in the XZW interaction is included by creating a series of graphs at different values of W. Thus the straightforward logic of plotting simple regression equations of Y on X at levels of a second variable Z from the two-predictor case generalizes directly to the three-predictor case in which simple regressions of Y on X are plotted at levels of two other variables.

To create Figure 4.1, values of X_L and X_H, one standard deviation below and above the mean of X, were substituted into each of the four simple regression equations of Table 4.1c. For example, for Z_H and W_H the simple slope equation from Table 4.1 is $\hat{Y} = 4.521 + 22.811$. Then with $s_X = 7.070$, the points on the graph for X_L and X_H are as follows:

$$\text{For } X_L = -7.070, \qquad \hat{Y} = 4.521(-7.070) + 22.881 = -9.082$$

$$\text{For } X_H = 7.070, \qquad \hat{Y} = 4.521(7.070) + 22.881 = 54.844$$

These two values are used to draw the regression line of Y on X at Z_H and W_H.

The reader should note that the analyst is not confined to the format of Figure 4.1. One might alternatively display the regression of Y on X at levels of W within each graph, with each graph confined to one level of Z. Or, one might plot the regression of Y on Z at levels of W within each graph, with each graph confined to one level of X. Plotting the interaction in various ways can often be useful in the interpretation of higher order interactions. However, theory may provide guidance in the organization of the plot. For example, in research on the relationship between life stress and health, life stress is typically seen as the "primary" independent variable whose effects may be modified by other variables (e.g., social support; perceived control over one's own health). Researchers would typically depict life stress on the X axis (abscissa) of the plot to emphasize the central importance of this variable.

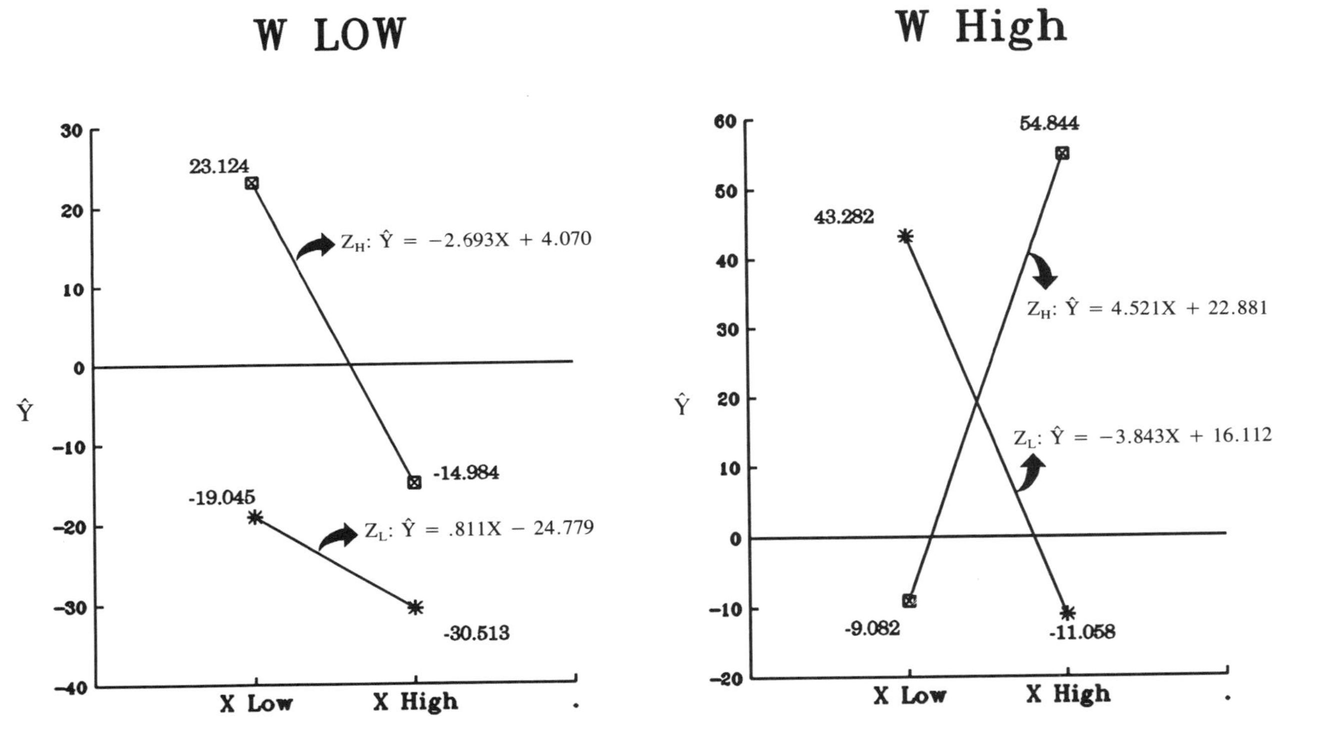

Figure 4.1. Three Predictor Linear *X* by Linear *Z* by Linear *W* Interaction

Testing Simple Slopes for Significance

Tests of simple slopes follow the same general proceedures as for testing the XZ interaction in Chapter 2. For the three-predictor interaction, the simple slope for the regression of Y on X is $(b_1 + b_4Z + b_5W + b_7ZW)$. The simple slope may be tested for significance at any combination of values of the continuous variables Z and W that are chosen. For readers familiar with ANOVA, these tests are analogous to tests of "simple simple main effects" (Winer, 1971).

The general expression for the standard error of the simple slope for Y on X at values of Z and W is as follows:[2]

$$s_b = [s_{11} + Z^2 s_{44} + W^2 s_{55} + Z^2 W^2 s_{77} + 2Zs_{14} + 2Ws_{15} + 2ZWs_{17} + 2ZWs_{45} + 2WZ^2 s_{47} + 2W^2 Zs_{57}]^{1/2} \quad (4.3)$$

Substituting appropriate values from $\mathbf{S}_b$ and varying values of Z and W into equation 4.3 yields the standard errors of the simple slopes.

Computation of the simple slope variance for the Z_H, W_H simple regression equation is shown for the three-predictor example in Table 4.2. As before, the standard error s_b becomes the denominator of the t-test for the simple slope, with $(n - k - 1)$ degrees of freedom, where k is the total number of predictors not including the regression constant, here 7.

We have limited our simple slope analysis to the values Z_H, Z_L and W_H, W_L, one standard deviation above and below Z and W, respectively. This is not required: Analysts are free to choose any combinations of Z and W that are meaningful in their own research area.

Standard Errors by Computer

The computer approach to finding simple slopes, standard errors, and t-values that we developed for the two-predictor case also generalizes directly to the three-predictor case. The three steps outlined in Chapter 2 are followed.

1. For both X and W, new variables Z_{CV} and W_{CV} are computed. These are Z and W minus the specific conditional values CV_Z and CV_W at which the regression of Y on X will be examined, that is, $Z_{CV} = Z - CV_Z$ and $W_{CV} = W - CV_W$. The conditional values typically would be Z_H, Z_L, W_H, and W_L.

2. For each pair of transformed values Z_{CV} and W_{CV}, the crossproducts of these terms with each other and with X are formed: $(X)(Z_{CV})$, $(X)(W_{CV})$, $(Z_{CV})(W_{CV})$, and $(X)(Z_{CV})(W_{CV})$.

Table 4.2
Variances, Standard Errors, and t-Tests for Simple Slopes in the Three-Predictor Interaction

a. $\mathbf{S}_b$: Variance–Covariance Matrix of b_s

	b_1	b_2	b_3	b_4	b_5	b_6	b_7
b_1	0.71498	−0.74621	−0.46138	0.00744	−0.02715	−0.06455	−0.01517
b_2	−0.74621	3.89690	−1.28826	−0.00452	−0.10133	−0.08607	−0.06742
b_3	−0.46138	−1.28826	33.54756	−0.01744	1.17015	−1.93137	−0.48024
b_4	0.00744	−0.00452	−0.01744	0.04979	−0.01563	−0.07925	−0.00338
b_5	0.02715	−0.10133	−1.17015	−0.01563	0.64088	−0.74581	−0.01669
b_6	−0.06455	−0.08607	−1.93137	−0.07925	−0.74581	3.40141	0.06244
b_7	−0.01517	−0.06742	−0.48024	−0.00338	−0.01669	0.06244	0.03954

b. Weight Vector for $Z_H = 3.096$ and $W_H = 1.045$

$$w' = [1 \quad 0 \quad 0 \quad 3.096 \quad 1.045 \quad 0 \quad (3.096)(1.045)]$$

c. Variance of Simple Slope of Y on X at Z_H and W_H (Using Expression 4.3)

$$\begin{aligned} s_b^2 = {} & 0.71498 + (3.096)^2(0.04979) + (1.045)^2(0.64088) + (3.096)^2(1.045)^2(0.03954) \\ & + 2(3.096)(0.00744) + 2(1.045)(0.02715) + 2(3.096)(1.045)(-0.01517) \\ & + 2(3.096)(1.045)(-0.01563) + 2(1.045)(3.096)^2(-0.00338) \\ & + 2(1.045)^2(3.096)(-0.01669) \end{aligned}$$

$$s_b^2 = 2.029 \qquad s_b = 1.424$$

d. Standard Errors and t-Tests for Simple Slopes of Y on X Shown in Figure 4.1

		Simple Slope	Standard Error	t-test
(i)	At Z_H, W_H	4.521	1.424	3.17**
(ii)	At Z_L, W_H	−3.843	1.600	−2.40*
(iii)	At Z_H, W_L	−2.693	1.565	−1.72+
(iv)	At Z_L, W_L	−0.811	1.478	−0.55

** $p < .01$; * $p < .05$; + $p < .10$

3. Taking each pair of Z_{CV} with W_{CV} (e.g., Z and W one standard deviation above their means) in turn, the criterion Y is regressed on X, Z_{CV}, W_{CV}, $(X)(Z_{CV})$, $(X)(W_{CV})$, $(Z_{CV})(W_{CV})$, and $(X)(Z_{CV})(W_{CV})$. The resulting regression coefficient b_1 for X is the simple regression coefficient of Y on X at the specific values Z_{CV} and W_{CV}. The t-test (using the reported standard error) of b_1 provides the test of the simple slope. The regression constant is that for the simple regression in question.

The computer analysis of the XZW interaction explored in Table 4.2 is given in Table 4.3. The overall regression analysis is given in Table 4.3a; note that predictors X, Z, and W are centered with $s_X = 7.070$, $s_Z =$

Table 4.3
Computation of Simple Slope Analysis by Computer for the *XZW* Interaction in the Regression Equation
$\hat{Y} = b_1X + b_2Z + b_3W + b_4XZ + b_5XW + b_6ZW + b_7XZW + b_0$

a. Overall Analysis with Centered *X*, *Z*, and *W*

(i) means and standard deviations

	Mean	Std Dev
Y	11.670	106.967
X	0.000	7.070
Z	0.000	3.096
W	0.000	1.045
XZ	11.299	23.598
XW	2.111	7.610
ZW	0.993	3.317
XZW	0.968	30.203

(ii) Regression analysis

Variable	B	SE B	T	Sig T
X	−0.706801	0.845566	−0.836	.4037
Z	2.876090	1.974057	1.457	.1459
W	14.283066	5.792025	2.466	.0141
XZ	0.523446	0.223133	2.346	.0195
XW	1.000706	0.800550	1.250	.2120
ZW	−1.706190	1.844291	−0.925	.3555
XZW	0.791742	0.198858	3.981	.0001
(Constant)	4.570963	5.668610	0.806	.4205

b. Computation of WABOVE, WBELOW, ZABOVE, ZBELOW and Crossproduct Terms Required for Simple Slope Analysis

```
COMPUTE WABOVE=W−1.045
COMPUTE WBELOW=W−(−1.045)
COMPUTE ZABOVE=Z−3.096
COMPUTE ZBELOW=Z−(−3.096)
COMPUTE XZB=X*ZBELOW
COMPUTE XZA=X*ZABOVE
COMPUTE XWB=X*WBELOW
COMPUTE XWA=X*WABOVE
COMPUTE ZBWB=ZBELOW*WBELOW
COMPUTE ZBWA=ZBELOW*WABOVE
COMPUTE ZAWB=ZABOVE*WBELOW
COMPUTE ZAWA=ZABOVE*WABOVE
COMPUTE XZBWB=X*ZBELOW*WBELOW
COMPUTE XZBWA=X*ZBELOW*WABOVE
COMPUTE XZAWB=X*ZABOVE*WBELOW
COMPUTE XZAWA=X*ZABOVE*WABOVE
```

Table 4.3, continued

c. Regression Analysis with ZABOVE, WABOVE, and Appropriate Crossproducts Yielding Simple Slope Analysis at Z_H and W_H (Regression of Y on X One Standard Deviation Above the Mean of Z and One Standard Deviation Above the Mean of W)

(i) Means and standard deviations

	Mean	Std Dev
Y	11.670	106.967
X	0.000	7.070
ZABOVE	−3.096	3.096
WABOVE	−1.045	1.045
XZA	11.299	32.225
XWA	2.111	10.749
ZAWA	4.229	6.004
XZAWA	−17.376	56.826

(ii) Regression Analysis

Variable	B	SE B	T	Sig T
X	4.521064	1.424386	3.174	.0016
ZABOVE	1.093121	2.726067	0.401	.6886
WABOVE	9.000700	7.361504	1.223	.2222
XZA	1.350816	0.293089	4.609	.0000
XWA	3.451939	0.957375	3.606	.0004
ZAWA	−1.706190	1.844291	−0.925	.3555
XZAWA	0.791742	0.198858	3.981	.0001
(Constant)	22.881068	10.730287	2.132	.0336

d. Regression Analysis with ZABOVE, WBELOW, and Appropriate Crossproducts Yielding Simple Slope Analysis at Z_H and W_L (Regression of Y on X One Standard Deviation Above the Mean of Z and One Standard Deviation Below the Mean of W)

(i) Means and standard deviations

	Mean	Std Dev
Y	11.670	106.967
X	0.000	7.070
ZABOVE	−3.096	3.096
WBELOW	1.045	1.045
XZA	11.299	32.225
XWB	2.111	10.462
ZAWB	−2.242	5.114
XZAWB	6.239	45.539

(ii) Regression analysis

Variable	B	SE B	T	Sig T
X	−2.693489	1.565075	−1.721	.0860
ZABOVE	4.659059	2.791273	1.669	.0959
WBELOW	9.000700	7.361504	1.223	.2222
XZA	−0.303924	0.316296	−0.961	.3372
XWB	3.451939	0.957375	3.606	.0004
ZAWB	−1.706190	1.844291	−0.925	.3555
XZAWB	0.791742	0.198858	3.981	.0001
(Constant)	4.069605	12.138283	0.335	.7376

3.096, and $s_W = 1.045$. Table 4.3b shows the computation of the new (transformed) variables from Step (1) above:

(a) WABOVE = $W - (1.045)$, for the regression of Y on X at $CV_W = 1.045$, one standard deviation above the mean of W;
(b) WBELOW = $W - (-1.045)$, for the regression of Y on X at $CV_W = -1.045$, one standard deviation below the mean of W;
(c) ZABOVE = $Z - (3.096)$, for the regression of Y on X at $CV_Z = 3.096$, one standard deviation above the mean of Z; and
(d) ZBELOW = $Z - (-3.096)$, for the regression of Y on X at $CV_Z = -3.096$, one standard deviation below the mean of Z.

Table 4.3b also shows the computation of crossproduct terms prescribed in Step (2) above. Table 4.3c provides the regression of Y on X at ZABOVE and WABOVE, that is, the regression of Y on X one standard deviation above the means of Z and W. The b_1 is the simple slope coefficient, also given in Table 4.1c, equation (i), and again in Table 4.2. The standard error of b_1 is the same value as was found using equation 4.3 (see Table 4.2); the t-test for b_1 is that for the simple slope. Finally, Table 4.3d provides the simple slope analysis one standard deviation above the mean of Z and one standard deviation below the mean of W, at ZABOVE and WBELOW.

Crossing Point of Simple Regression Equations with Three-Predictor Interaction

We showed in Chapter 2 for the two-predictor regression equation $\hat{Y} = b_1X + b_2Z + b_3XZ + b_0$ that the simple regression equations of Y on X would cross at the value $X_{\text{cross}} = -b_2/b_3$. In the case in which the b_3 coefficient for the interaction is zero, the simple regression equations do not cross, that is, they are parallel and the value of X_{cross} is undefined.

We can easily extend this analysis and our criterion for ordinal versus disordinal status to the three-predictor case. Consider two simple regression equations of Y on X at a fixed value of W but at differing values of Z, say Z_L and Z_H. We substitute into equation 4.1 first the value of Z_L (expression 1) and then the value of Z_H (expression 2). The two expressions are then set equal to one another, yielding the value of X the point at which the two simple regression lines cross:

$$b_1X + b_2Z_L + b_3W + b_4XZ_L + b_5XW + b_6Z_LW + b_7XZ_LW + b_0$$

$$= b_1X + b_2Z_H + b_3W + b_4XZ_H + b_5XW + b_6Z_HW + b_7XZ_HW + b_0$$

The value of X is:

$$X_{\text{cross}} \text{ at specified } W = \frac{-(b_2 + b_6 W)}{(b_4 + b_7 W)} \tag{4.4}$$

Note that the denominator of this expression will be zero when both the XZ and the XZW interactions do not exist, that is, when b_4 and b_7 are zero. When the denominator is zero, the simple regression lines are parallel. The fact that W appears in equation 4.4 indicates that the crossing point for the simple regressions of Y on X at values of Z depend on the specific value of W. Figure 4.1 illustrates this point nicely; at "W Low," the left-hand graph of Figure 4.1, the simple regression lines do not cross within the range of X that is portrayed; at "W High," they do cross. Now instead of a single cross-over point, there is a line of cross-over points, with each point on the line corresponding to a different value of W.

If equation 4.4 is evaluated at $W = 1.045$ corresponding to the "W High" graph of Figure 4.1, then

$$X_{\text{cross}} \text{ at } W_{\text{H}} = \frac{-[2.8761 + (-1.7062)(1.045)]}{[0.5234 + (0.7917)(1.045)]}$$

$$= -0.8093$$

For $W_{\text{L}} = -1.045$ corresponding to the "W Low" graph of Figure 4.1, X_{cross} at $W_{\text{L}} = 15.3310$. Note that at W_{H}, the crossing point is well within one standard deviation of the mean ($s_X = 7.07$), whereas for W_{L}, the crossing point is over two standard deviations above the mean.

Alternatively, if one has plotted regression of Y on X at levels of W within separate graphs confined to particular levels of Z, then the X value of interest is

$$X_{\text{cross}} \text{ at specified } Z = \frac{-(b_3 + b_6 Z)}{(b_5 + b_7 Z)} \tag{4.5}$$

The denominator of this expression will be zero if both the XW and XZW interactions are zero.

Simple Slopes and their Variances in a Series of Regression Equations

The simple slope and variance expressions for equations of similar form follow orderly patterns. The regression of Y on X for equation 2.1 is pre-

Table 4.4
Expressions for Simple Slopes and Their Variances for Various Regression Equations Containing Two- and Three-Predictor Interactions

Case	Equation	Regression	Simple Slope	Variance of Simple Slope (s_b^2)
(1a)	$\hat{Y} = b_1X + b_2Z + b_3XZ + b_0$	Y on X	$(b_1 + b_3Z)$	$s_{11} + 2Zs_{13} + Z^2s_{33}$
(1b)	$\hat{Y} = b_1X + b_2Z + b_3XZ + b_0$	Y on Z	$(b_2 + b_3X)$	$s_{22} + 2Xs_{23} + X^2s_{33}$
(2)	$\hat{Y} = b_1X + b_2Z + b_3W + b_4XZ + b_6ZW + b_0$	Y on X	$(b_1 + b_4Z)$	$s_{11} + 2Zs_{14} + Z^2s_{44}$
(3)	$\hat{Y} = b_1X + b_2Z + b_3W + b_4XZ + b_5XW + b_6ZW + b_0$	Y on X	$(b_1 + b_4Z + b_5W)$	$s_{11} + Z^2s_{44} + W^2s_{55} + 2Zs_{14} + 2Ws_{15} + ZWs_{45}$
(4)	$\hat{Y} = b_1X + b_2Z + b_3W + b_4XZ + b_5XW + b_6ZW + b_7XZW + b_0$	Y on X	$(b_1 + b_4Z + b_5W + b_7ZW)$	$s_{11} + Z^2s_{44} + W^2s_{55} + Z^2W^2s_{77} + 2Zs_{14} + 2Ws_{15} + 2ZWs_{17} + 2ZWs_{45} + 2WZ^2s_{47} + 2W^2Zs_{57}$

NOTE: In Case 2, there is no b_5 coefficient in order that the progression of Cases 2–4 have consistent notation.

sented as Case 1a in Table 4.4. Case 2 of Table 4.4 presents the regression of Y on X in an equation with one level of increase in complexity over Case 1: The addition of a W first order term and a ZW interaction. Because neither of these new terms involves X, they have no effect on the expressions for the simple slope of Y on X or its variance. The simple slopes for Case 1a and Case 2 have the same structure. Case 3 of Table 4.4 is increased in complexity from Case 2 by the addition of the XW interaction; it contains all three two-way interactions among X, Z, and W, but not the three-way interaction. Since the XW interaction, tested with the b_5 coefficient, does contain X, the b_5 coefficient appears in the simple slope expression for Y on X; the variance of b_5 and its covariance with other predictors appear in the expression for the variance of the simple slope. Finally, Case 4 is the complete equation including the XZW interaction, to which Chapter 4 is devoted. The reader may follow the patterns illustrated in Table 4.4 to generate simple slopes and variance expressions for equations of varying complexity but which contain only linear terms and products of linear terms.

Summary

The procedures for specifying, testing, and interpreting a three-way XZW interaction are shown to be straightforward generalizations of the procedures developed for the two-way XZ interaction in Chapter 2. Plotting the three-way interaction is accomplished by plotting two-way interactions at values of the third variable, just as in ANOVA with three-factor interactions. The methods for generating tests of simple slopes are generalized to the three-predictor interaction, and the computer approach to these tests is illustrated for the three-predictor case. General patterns in the structure of simple slopes and their variances are illustrated for regression equations of increasing complexity.

Notes

1. Social science research areas differ in their position about the permissibility of omitting lower order terms in regression equations. The only case in which a justification for this practice may be offered is when strong theory dictates a lower order effect must equal zero (see Fisher, 1988; Kmenta, 1986).

2. The weight vector used to generate this expression is $\mathbf{w}' = [1\ 0\ 0\ Z\ W\ 0\ ZW]$ and $\mathbf{S}_b$ is a 7×7 matrix. Then, $s_b^2 = \mathbf{w}'\mathbf{S}_b\mathbf{w}$ as in equation 2.8.

5 Structuring Regression Equations to Reflect Higher Order Relationships

Many cases exist in the social sciences in which complex relationships are expected between predictors and a criterion. These more complex relationships often take the form of a monotonically increasing (or decreasing) curvilinear relationship or a U-shaped or inverted U-shaped function. For example, in psychology the well-known Yerkes–Dodson law (Yerkes & Dodson, 1908) predicts that the relationship between physiological arousal and performance will follow an inverted U-shaped function. To examine these relationships specific higher order terms must deliberately be built into the regression equation. If these higher order terms are omitted, nonlinearity will not be detected even when it does exist. Otherwise stated, the use of the regression equation $\hat{Y} = b_1X + b_2Z + b_3XZ + b_0$ makes the assumption that only linear relationships of predictors to criterion and a linear by linear interaction of the general form depicted in Figure 2.1 potentially exist. The analogous situation in ANOVA is the use of only two levels of each of two factors, a procedure that makes the implicit assumption that the relationship between each factor and the criterion can only be linear and that any possible interaction must be linear by linear in form.

In this chapter we explore how higher order relationships are represented and tested in MR. The reader should be aware at the outset that this chapter is relatively more complex than the previous ones; hence it may go more slowly. To help the reader, we have taken a two-stage approach to our presentation. First, we have illustrated what regression equations containing higher order terms look like in terms of the forms of relationships they represent. We begin with a relatively simple equation (Case 1 below) and gradually build in complexity (Cases 2, 3, and 4). Second, we consider the post hoc probing of equations presented in the first stage, again beginning with Case 1 and working through Cases 2, 3, and 4. The case numbers we have used throughout the chapter are the same as those in Table 5.1 for each equation. Table 5.1 summarizes the cases we will consider in depth here.

Our discussion of the representation of curvilinear relationships is limited to terms involving no more than second powers of the predictor variables. Certainly effects represented by still higher order powers may occur and our prescriptions may be generalized to these relationships as well. However, at present, relationships with higher than second power terms are rarely, if ever, hypothesized to exist in the social sciences.

Structuring and Interpreting Regression Equations Involving Higher Order Relationships

For purposes of clarity, throughout this section we will use *effect* to signify a general source of variance that parallels the typical partition in ANOVA (e.g., a main effect, an interaction) and *component* to signify a single predictor term (e.g., X^2Z) that is part of the effect. We will again assume that the predictor variables have been centered to facilitate interpretation of the regression coefficients.

Case 1: Curvilinear X Relationship

Suppose that we expect a single predictor X to have a curvilinear relationship with Y. In this case we would use the following regression equation to represent the relationship:

$$\hat{Y} = b_1X + b_2X^2 + b_0 \tag{5.1}$$

The X and X^2 terms represent the linear and quadratic *components* of the overall "main" *effect* of X, each with one degree of freedom. Note that both the X and the X^2 terms must be included in the equation, even if it is expected that there is only a quadratic relationship[1] between X and Y. As is illustrated in Figure 5.1, this equation fits many different appear-

Table 5.1
Expressions for Simple Slopes and Their Variances for Various Regression Equations Containing Second-Order Terms

Case	Equation	Regression	Simple Slope	Variance of Simple Slope (s_b^2)
(1)	$\hat{Y} = b_1X + b_2X^2 + b_0$	Y on X	$(b_1 + 2b_2X)$	$s_{11} + 4Xs_{12} + 4X^2s_{22}$
(2)	$\hat{Y} = b_1X + b_2X^2 + b_3Z + b_0$	Y on X	$(b_1 + 2b_2X)$	$s_{11} + 4Xs_{12} + 4X^2s_{22}$
(3a)	$\hat{Y} = b_1X + b_2X^2 + b_3Z + b_4XZ + b_0$	Y on X	$(b_1 + 2b_2X + b_4Z)$	$s_{11} + 4X^2s_{22} + Z^2s_{44} + 4Xs_{12} + 2Zs_{14} + 4XZs_{24}$
(3b)	$\hat{Y} = b_1X + b_2X^2 + b_3Z + b_4XZ + b_0$	Y on Z	$(b_3 + b_4X)$	$s_{33} + 2Xs_{34} + X^2s_{44}$
(4a)	$\hat{Y} = b_1X + b_2X^2 + b_3Z + b_4XZ + b_5X^2Z + b_0$	Y on X	$(b_1 + 2b_2X + b_4Z + 2b_5XZ)$	$s_{11} + 4X^2s_{22} + Z^2s_{44} + 4X^2Z^2s_{55} + 4Xs_{12} + 2Zs_{14} + 4XZs_{24} + 4XZs_{15} + 8X^2Zs_{25} + 4XZ^2s_{45}$
(4b)	$\hat{Y} = b_1X + b_2X^2 + b_3Z + b_4XZ + b_5X^2Z + b_0$	Y on Z	$(b_3 + b_4X + b_5X^2)$	$s_{33} + X^2s_{44} + X^4s_{55} + 2Xs_{34} + 2X^2s_{35} + 2X^3s_{45}$

ances of the relationship between X and Y. Because the X and X^2 terms form a building block for more complex equations, the interpretation of the regression coefficients in equation 5.1 is explored here in some detail.

With centered predictors, the b_1 coefficient indicates the overall linear trend (positive or negative) in the relationship between X and Y across the observed data. If the linear trend is predominantly positive, as in Figures 5.1a,b, b_1 is positive; if the linear trend is predominantly negative, as in Figure 5.1c, then b_1 is negative. For the completely symmetric U-shaped and inverted U-shaped relationships depicted in Figures 5.1d and 5.1e respectively, b_1 is zero. The interpretation of b_1, then, is consistent with all previous interpretations in Chapters 2 through 4, when centered predictors are employed.

The b_2 coefficient indicates the direction of curvature. If the relationship is concave upward, as in Figures 5.1a,d, then b_2 is positive; if the relationship is concave downward, as in Figures 5.1b,c,e, then b_2 is negative. When the curve is concave upward (b_2 positive), we often are interested in the value of X at which $\hat{Y}$ takes on its lowest value, the *minimum* of the curve, as in Figures 5.1a,d. When the curve is concave downward (b_2 negative), we may seek the value of X at which $\hat{Y}$ reaches its highest value, the *maximum* of the curve, as in Figures 5.1b,c,e. As we explain later, the maximum or minimum point of the function is reached when $X = -b_1/2b_2$. If this value falls within the meaningful range of X, then the relationship is nonmonotonic and may appear as in Figure 5.1d,e. If this value falls outside the meaningful range of the data, then the relationship appears monotonic, as in Figures 5.1a,b,c. It should be noted that the distance from the maximum or minimum point to the mean of X is invariant under additive transformation.

When X bears a linear relationship to Y (i.e., no higher order term containing X appears in the equation), a one-unit change in X is associated with a one-unit change in Y. In contrast, when X bears a curvilinear relationship to Y (i.e., there is a higher order term containing a power of X in the equation such as X^2), then the change in Y for a one-unit change in X depends upon the value of X. This can be verified by noting the change in Y as a function of X in Figure 5.1.

A Progression of More Complex Equations with Curvilinear Relationships

We now consider in turn each of a series of regression equations that are used to represent specific higher order relationships. To illustrate our

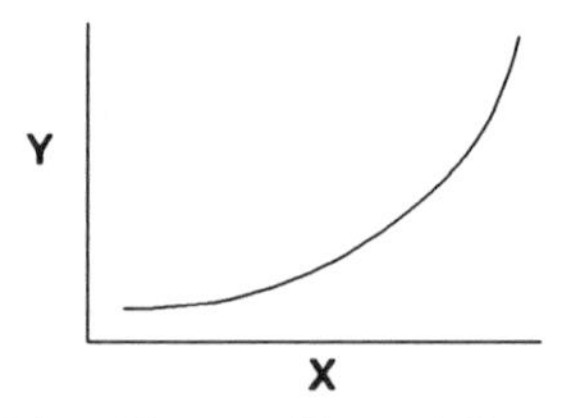

a. A Predominantly Positive, Concave Upward Curve (b_1 Positive, b_2 Positive)

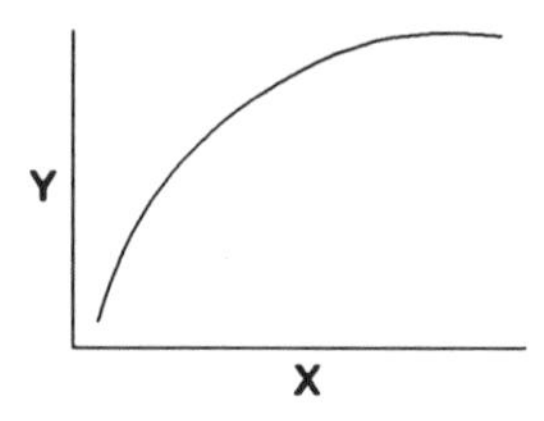

b. A Predominantly Positive, Concave Downward Curve (b_1 Positive, b_2 Negative)

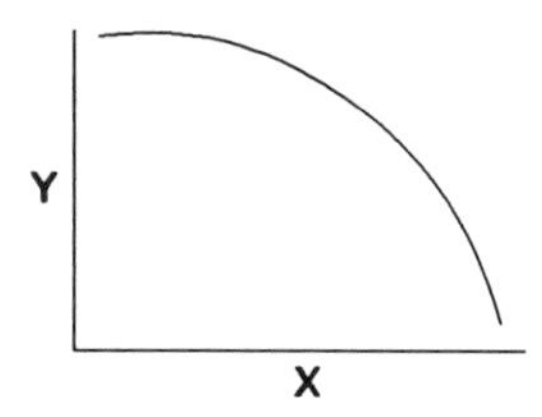

c. A Predominantly Negative, Concave Downward Curve (b_1 Negative, b_2 Negative)

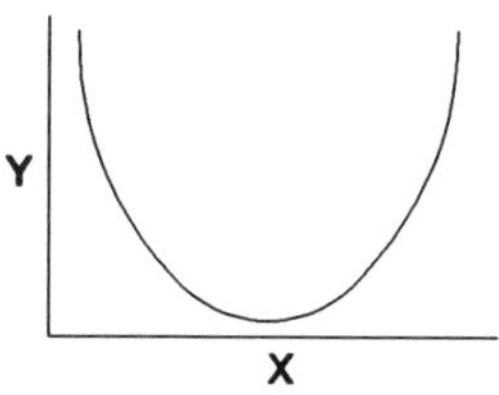

d. U-Shaped Function ($b_1 = 0$, b_2 Positive)

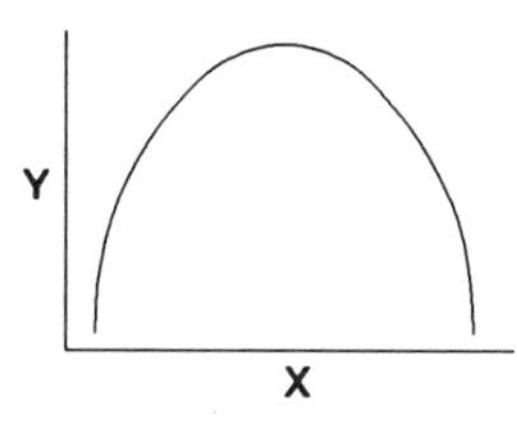

e. An Inverted U-Shaped Function ($b_1 = 0$, b_2 Negative)

Figure 5.1. Some Forms of the Equation: $\hat{Y} = b_1X + b_2X^2 + b_0$

prescriptions for testing and post hoc probing of interactions, we will employ throughout this chapter a single simulated data set. The same 400 bivariate normal pairs of scores used in Chapter 2 were employed here. X and Z were centered ($\bar{X} = 0$; $\bar{Z} = 0$) and they are moderately correlated, $r_{XZ} = 0.42$. The higher order terms X^2, XZ, and X^2Z were then generated from the centered terms, and all terms were used to produce $\hat{Y}$ using the regression equation $\hat{Y} = 0.2X + 5.0X^2 + 2.0Z + 2.5XZ + 1.5X^2Z$. Finally, observed Y scores were generated from predicted $\hat{Y}$ scores by the addition of normally distributed random error. Linking this simulation to a substantive example, predictor X represents an individual's self-concept (i.e., how well or poorly an individual evaluates himself or herself overall). The criterion Y represents an individual's level of self-disclosure, or the extent to which the individual shares personal information with others. Self-disclosure (Y) has been found to be a U-shaped function of self-concept (X); individuals with low or high self-concepts tend to disclose more about themselves than persons with moderate self-concepts. Predictor Z represents the amount of alcohol consumed in a social situation in which an individual has an opportunity to self-disclose. Self-disclosure is expected to increase with increased alcohol consumption; a linear relationship is assumed in the absence of theory specifying a more complex relationship.

Case 2: Curvilinear X Relationship and Linear Z Relationship

In a two predictor equation, if it is expected that predictor Z will have a linear effect on the outcome Y but that predictor X may have a curvilinear effect on Y, then the following equation would be used:

$$\hat{Y} = b_1X + b_2X^2 + b_3Z + b_0 \tag{5.2}$$

Figure 5.2a illustrates an equation of this form. The simple regression lines are plotted in two different ways: First, the simple regression lines of Y on X at values of Z are illustrated in Figure 5.2a(1). Each curve represents a single level of alcohol consumption (Z_L, Z_M, Z_H). The same curvilinear relationship of Y to X is observed at each level of alcohol consumption. The simple regression lines of Y on Z at values of X are portrayed in Figure 5.2a(2). Note that the regression of Y on X is curvilinear, whereas the regression of Y on Z is linear. In both cases, the simple regression lines are parallel, because there are no terms to represent an interaction between X and Z. Because the relationship of Z to Y is linear, there is equal displacement of the simple regression lines of Y on X at levels of Z across the range of Z; a constant change in Z, from Z_L to Z_M or from Z_M to Z_H produces a constant change in Y. However, the curvilinear relationship between X and Y means that a one-unit change in X does not produce a constant change in Y across the continuum. This curvilinearity gives rise to the unequal displacements of the simple regression lines of Y on Z at levels of X depicted in Figure 5.2a(2).

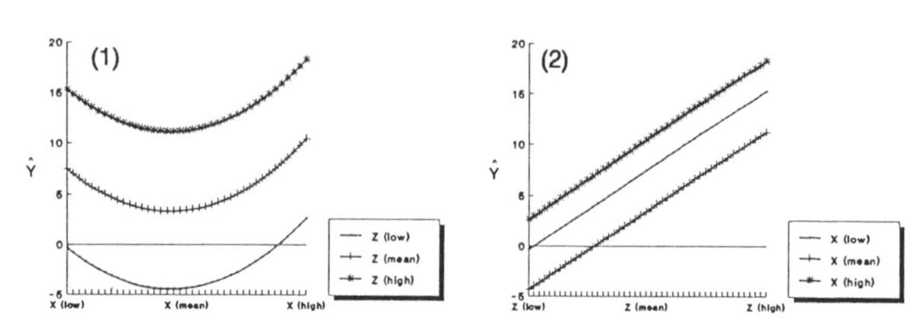

a. Relationships Represented in Equation $\hat{Y} = 1.59X + 6.18X^2 + 3.55Z + 3.44$

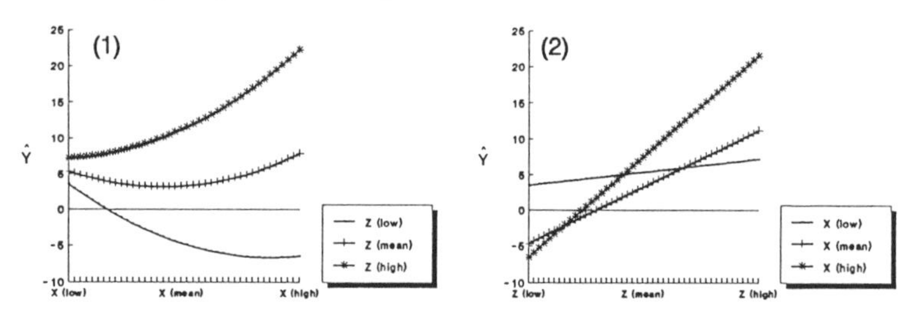

b. Relationships Represented in Equation $\hat{Y} = 1.13X + 3.56X^2 + 3.61Z + 2.93XZ + 3.25$

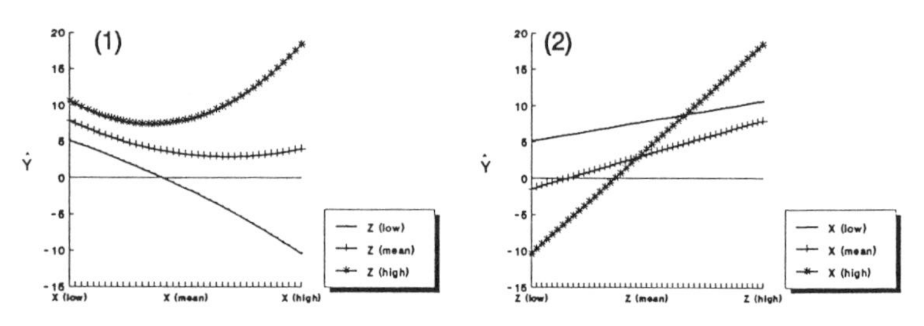

c. Relationships Represented in Equation $\hat{Y} = -2.04X + 3.00X^2 + 2.14Z + 2.79XZ + 1.96X^2Z + 3.50$

Figure 5.2. Relationships Represented by Equations Containing Higher Order Terms

Case 3: Curvilinear X Relationship,
Linear Z Relationship, and
Linear by Linear XZ Interaction

We now add a term representing a simple linear by linear interaction to the previously considered case, resulting in the following equation:

$$\hat{Y} = b_1X + b_2X^2 + b_3Z + b_4XZ + b_0 \qquad (5.3)$$

In our example we might hypothesize that the curvilinear relationship of self-disclosure to self-concept exists at all levels of alcohol consumption, but that alcohol consumption interacts with self-concept in determining self-disclosure. At high alcohol consuumption, the high self-disclosure of those with high self-concepts would even be increased (a bit of self-aggrandizement, perhaps), whereas at a very low alcohol consumption level, the self-disclosure of those with high self-concepts would be diminished (a bit of humility, perhaps). This is a complex hypothesis, but not out of the realm of possibility.

The effect of the introduction of the linear by linear interaction component is illustrated in Figure 5.2b for the simulated data set. Again, the regressions are plotted in two ways. Figure 5.2b(1) shows the simple regression lines depicting the curvilinear regression of Y on X at levels of Z. Because of the interaction, the curves are no longer parallel. However, since the interaction only involves first order terms in both X and Z, the curves are identical in shape, that is, all are mildly concave upward. But, in keeping with the hypotheses of our example, at Z_H there is an overall increasing trend; while at Z_L the trend is an overall decrease in self-disclosure. Figure 5.2b(2) shows the same interaction from the perspective of the simple regression lines depicting the regression of Y on Z at levels of X. Note that the simple regression lines no longer cross at one single point, as they did in equation 2.1 (see also Figure 3.1).

Case 4: Curvilinear X Relationship,
Linear Z Relationship, and
a Curvilinear X by Linear Z Interaction

A quadratic X by linear Z interaction component is now introduced:

$$\hat{Y} = b_1X + b_2X^2 + b_3Z + b_4XZ + b_5X^2Z + b_0 \qquad (5.4)$$

This equation is again plotted in two ways in Figure 5.2c, illustrating first the regression of Y on X and then the regression of Y on Z. As is most clearly seen in Figure 5.2c(1), the meaning of the quadratic by linear X^2Z term is that the quadratic relationship between X and Y varies in form as a function of the value of Z. If we had hypothesized that the curvilinear relationship between self-disclosure and self-concept would be increasingly manifested as alcohol consumption increased, equation 5.4 would have been appropriate to test the prediction.

Case 5: Curvilinear X Relationship,
Curvilinear Z Relationship,
and Their Interactions

Suppose finally that both the X and Z predictors were expected to have curvilinear effects. In this case, both predictors would be treated as predictor X was in equation 5.2 as represented in the first four terms of equation 5.5. In addition, up to four components of the interaction may be included in the regression equation, namely the XZ, XZ^2, X^2Z, and X^2Z^2 terms:

$$\hat{Y} = b_1X + b_2Z + b_3X^2 + b_4Z^2 + b_5XZ + b_6XZ^2 + b_7X^2Z + b_8X^2Z^2 + b_0 \qquad (5.5)$$

The two plots of this equation would yield graphs like that in Figure 5.2c(1) for both the regression of Y on X at levels of Z and the regression of Y on Z at levels of X. (We mention this case but do not consider it further.)

Representation of Curvilinearity in ANOVA Versus MR

In MR two separate predictors are required to represent a "main effect" that involves a second order term, as is shown in equation 5.1 To represent a two-way interaction, up to four additional terms may be required, as in equation 5.5. In contrast, in ANOVA we are accustomed to having one source of variation for each main effect and one source of variation for each interaction. Depending on the number of levels of each factor, the source of variation will have one or more than one degree of freedom. Regardless of the number of degrees of freedom associated with each main effect and interaction, each source of variation is separately tested for significance with a single omnibus test.

The key to understanding the direct equivalence of MR and ANOVA lies in the fact that each of the overall main effects and interactions in ANOVA aggregates together all the linear and higher order nonlinear partitions of variation that are represented by separate components in MR. Consider a main effect in ANOVA with three ordered levels of a factor and therefore two degrees of freedom. The main effect partition combines all of the linear and second order variation into one source of variation. If we follow the same procedure with MR equation 5.1, we find that the total predictable variation based on both of the predictors and two degrees of freedom is identical to the main effect variation of the ANOVA. In our example, suppose an experiment were performed with three groups of subjects, one with low self-concept, one with moderate self-concept, and one with high self-concept. In an ANOVA there would be one main effect of self-concept with two degrees of freedom that subsumed any linear trend in the relationship of self-concept to self-disclosure as well as the curvilinear relationship. (Note that we do not advocate splitting a continuous variable into groups so that ANOVA may be used—this is only for illustrative purposes. See Chapter 8 for a discussion of the costs of splitting continuous variables to create groups for an ANOVA.)

Alternatively, the main effect of the ANOVA can be partitioned into linear and quadratic components, each with one degree of freedom, using orthogonal polynomials (e.g., Kirk, 1982; Winer, 1971). Such equivalence of MR and ANOVA can also be shown for more complex cases such as equation 5.4. Here we would have a two-way ANOVA with three levels of Factor X and two levels of Factor Z. Now, in addition to the main effect of X with two degrees of freedom described above, there is an interaction with two degrees of freedom. This interaction effect corresponds to the two interaction terms in equation 5.4. Once again the overall ANOVA interaction could be partitioned into linear by linear and a quadratic by linear components with orthogonal polynomials.

Note, however, that there is an important difference between ANOVA and MR in usual practice. In ANOVA with multiple levels of a factor and the use of usual approaches to variance partitioning, any curvilinear variation is automatically subsumed in the variance partitions. In contrast, in MR the analyst specifically decides which terms need to be included: Terms to represent curvilinear relationships must be built systematically into the equation. This is not to say ANOVA and MR differ mathematically. Rather it is to say that the conventional partitions of variance operationalized in common statistical packages for ANOVA are structured so that all components of an effect are subsumed in the omnibus term for that effect; in MR, the structuring of the components of each effect is left entirely to the analyst.

Failure to structure the regression equation to reflect the theoretically expected curvilinear relationships and their interactions can lead to errors of interpretation, sometimes of substantial magnitude. For example, the relationships in Figures 5.2a,b,c are all based on the same data set. What differs between these portions of Figure 5.2 is the complexity of the equations used to fit the data. Recall that these data were simulated: We know that the data were in fact generated by an equation of the form represented in Figure 5.2c.

It may appear that the researcher must be more informed about the nature of the relationships of predictors to criteria to use MR than ANOVA. After all, the ANOVA automatically includes curvilinear components in the variance partitions. In actuality, in the planning of an experiment, the researcher must pick the number of levels of each factor based on some assumption or knowledge of the relationship of the factors to the criterion. If there is only a linear relationship expected, two levels suffice to estimate the linear effect (but three levels are required to test for nonlinearity). If curvilinear relationships are expected, at least three levels are required. In an analogous manner, the researcher using MR who is suspicious that there are nonlinearities in the relationships may explore these relationships by using equations containing appropriate higher order terms (see also Chapter 9).[2]

Post Hoc Probing of More Complex Regression Equations

The procedures for probing interactions of the form XZ and XZW developed in previous chapters directly generalize to more complex regression equations. In this section we show how simple regression equations can be derived and tested for the curvilinear X case and for each of the progression of regression equations just considered. Indeed, as a useful summary, expressions for the simple slopes and variances for each of these equations are summarized in Table 5.1. We also show how to calculate the crossing points for the simple regression lines corresponding to several of the equations. Finally, we introduce new tools for probing simple regression lines that are curvilinear in form.

Case 1: Curvilinear X Equation

We begin with the simple curvilinear (second order) equation involving only X and X^2 terms represented by equation 5.1, which is reproduced below:

$$\hat{Y} = b_1X + b_2X^2 + b_0$$

As before, we rewrite the equation to show the regression of Y on X:

$$\hat{Y} = (b_1 + b_2X)X + b_0 \tag{5.6}$$

From equation 5.6 we see that the regression of Y on X *depends* upon the specific value of X; that is, at any particular value of X, the value of Y may be decreasing, not changing much at all, or increasing. Inspection of Figure 5.3 also confirms this result.

The simple slope for the regression of Y on X in equation 5.1 is actually not $(b_1 + b_2X)$. Rather the correct expression for the simple slope is $(b_1 + 2b_2X)$. The straightforward approach of rearranging regression equations to identify simple slopes does *not* generalize to equations containing curvilinear components (e.g., X^2) of first order terms. Finding simple slopes for equations containing curvilinear components requires a bit of calculus, which we provide below, with graphic illustration. We strongly encourage readers to work through this presentation and the numerical example in order to understand the remainder of the chapter.

Simple Slope Expressions as Derivatives of Regression Equations

In considering the XZ and XZW interactions in previous chapters, we have only dealt with linear changes in Y associated with linear changes in X. But, how does one measure the regression of Y on X at a single value of X along a curve? A mathematical operation in calculus known as differentiation (taking the first derivative) provides the answer. This operation provides the slope of a tangent line to a curve at any point on the curve.[3] Figure 5.3a illustrates a tangent line to a curve at one value of X, namely X_i; the curve follows the form of equation 5.1. The tangent line shows the regression of Y on X at X_i. The slope of the tangent line to the curve at X_i measures the simple regression of Y on X at that specific value of X; thus the slope of the tangent line is the simple slope for the regression of Y on X at one value of X. Putting this all together *the first derivative of the curve with respect to X evaluated at one value of X is the simple slope of Y on X at that value of X.*

Using rules from calculus for differentiating a function leads to the first derivative of equation 5.1 with respect to X:

$$\frac{d\hat{Y}}{dX} = b_1 + 2b_2X \tag{5.7}$$

This derivative is the general expression for the slope of a tangent line to the curve, where the curve is defined by overall regression equation 5.1.

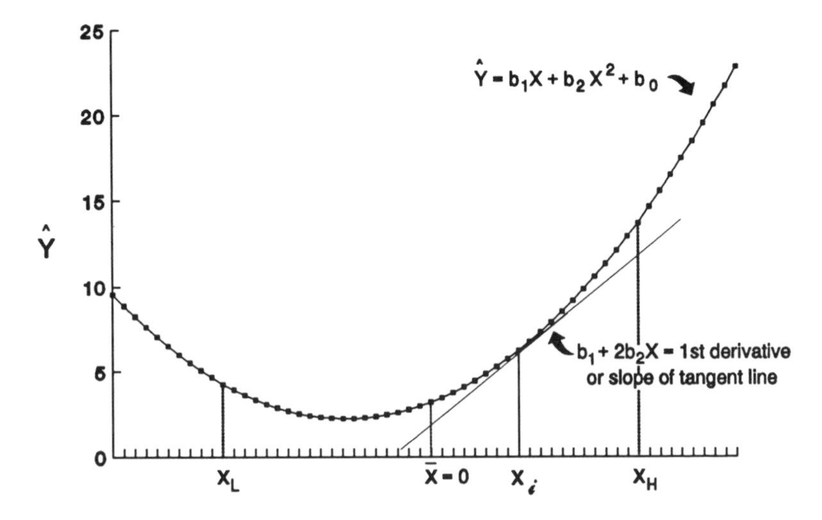

a. The Regression Equation (Curve) and the First Derivative (Tangent Line) at a Particular Value of X

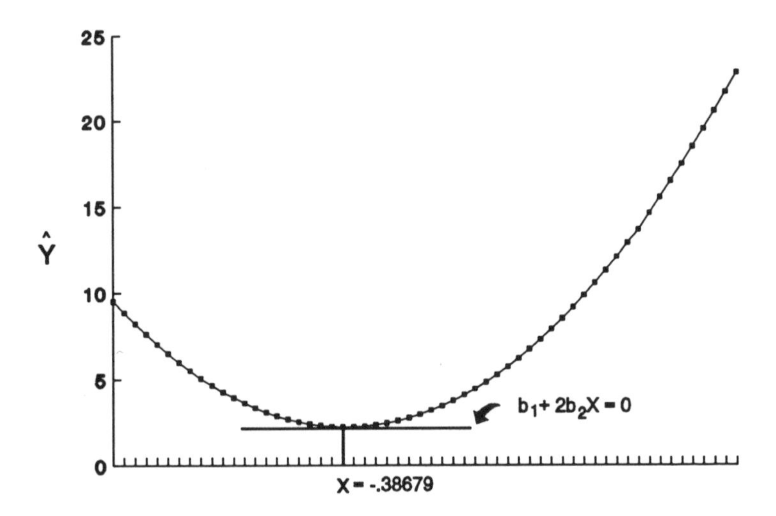

b. First Derivative Equals Zero at Minimum of Curve

Figure 5.3. The Regression of Y on X at Values of X in the Regression Equation: $\hat{Y} = 4.99X + 6.45X^2 + 3.20$

If the value X_i is substituted into equation 5.7, the resulting value is the value of the simple slope of Y on X at X_i, as illustrated in Figure 5.3a. Otherwise stated, equation 5.7 is an expression for the linear change in Y associated with a linear change in X at any specific value of X.

The general definition of the simple slope as the first (partial) derivative is applicable to all the simple slope expressions in Chapters 2 through 4. Consider the regression equation with one XZ interaction, $\hat{Y} = b_1X + b_2Z + b_3XZ + b_0$. We take the first (partial) derivative of this equation with respect to X:

$$\frac{\partial \hat{Y}}{\partial X} = b_1 + b_3Z \tag{5.8}$$

This is the same expression for the simple slope as was given in Chapter 2 and summarized in Table 4.4, Case 1.

At this point we will formalize the definition of simple slopes to encompass all the cases throughout the text. *The simple slope of the regression of Y on X is the first (partial) derivative of the overall regression equation with respect to the variable X.* This function indicates the slope of the regression of Y on X at a particular value of at least one variable, either some other variable Z, the variable X itself, or some combination of variables.

Returning to the simple curvilinear equation, what does its first derivative tell us? Consider the regression equation illustrated in Figure 5.3a: $\hat{Y} = 4.993X + 6.454X^2 + 3.197$, where X is centered. Using the expression for the derivative, we see that at the mean of X ($\bar{X} = 0$), the regression of Y on X is positive:

$$b_1 + 2b_2X = 4.993 + 2(6.454)(0) = 4.993$$

At X_H, one standard deviation above the mean ($s_X = 0.945$), $X_H = 0.945$ so that the regression of Y on X at X_H is $4.993 + 2(6.454)(0.945) = 17.191$. At $X_L = -0.945$, the regression of Y on X is -7.2051. Thus at one standard deviation below the mean (X_L), there is a negative relationship of X to Y; whereas at one standard deviation above the mean (X_H) the relationship is strongly positive.

Minimum or Maximum of Curve

At what point is the regression of Y on X equal to zero? Inspection of Figure 5.3b shows that the regression is zero at the minimum of the curve, that is, the value of X at which the predicted value of Y is its lowest. This point on the curve is where the tangent line to the curve (first derivative) has a slope equal to zero. Using expression 5.7, we solve for that value of X that causes expression 5.7 to equal zero:

$$b_1 + 2b_2X = 0$$

$$b_1 = -2b_2X$$

$$\frac{-b_1}{2b_2} = X \text{ at } Y_{\text{minimum}} \tag{5.9}$$

For these data, $X = -4.993/(2)(6.454) = -0.387$. At the value of $X = -0.387$, the regression of Y on X is zero and the predicted value of Y is minimum. In our example, this would indicate that self-disclosure is lowest just below the mean self-concept in the sample.

Note that for simple curvilinear regression equations representing inverted U-shaped functions the regression of Y on X is zero at the *maximum* predicted value of Y. Applied to curves such as those of Figure 5.1b,c,e equation 5.9 would yield the value of X at that maximum.[4]

Overall Trend in the Relationship of X to Y

At this point we revisit the b_1 coefficient in the overall regression equation 5.1 and contrast it with the simple slope of Y on X at any value of X. The b_1 coefficient, evaluated when centered X is equal to 0 (at the mean), represents the regression of Y on X at the mean of X, or the *conditional effect* at its mean. If it is positive, this indicates that in the region of the mean, Y is increasing; if negative, that Y is decreasing. For our example, because $b_1 = 4.993$ in Figure 5.3b, in the region of the mean of self-concept, self-disclosure increases as self-concept increases. Further, if one calculated the regression of Y on X for the value of X of each case in the observed sample (giving each case equal weight), on *average* this regression would be positive, with an average slope of 4.993. Across the observed cases, self-disclosure is on average higher with high than with low self-concept. In contrast to the b_1 coefficient, the simple slope, as before, measures the regression of Y on X at a single value of X. Evaluated at $X = 0$ for centered X, the simple slope will equal the b_1 coefficient, as has been true for the linear cases previously presented in Chapters 2 through 4.

Testing Simple Slopes

The computation of simple slopes to characterize higher order relationships provides a useful way to summarize these relationships and to place numerous results in context. The presence of curvilinear relationships, for example, has at times produced great consternation in research literature. Some researchers report positive relationships of X to Y; others find negative relationships; and still others report no relationship whatever. The basis of these discrepancies may well be in the range of X sampled across studies. Returning to Figure 5.3, one can see that if a small range of X at the low end of the continuum were sampled, a negative relationship would be observed. If a small range at the high end were sampled, a steeply positive relationship would be reported. Untangling these apparently conflicting results requires determining the range of values of X for which the relationship is negative, not different from zero, or positive. Simple slope tests aid in this task.

We may wish to test whether simple slopes representing the regression of Y on X at each value of X are significantly different from zero.[5] The standard error of the simple slope $(b_1 + 2b_2X)$ is

$$s_b = \sqrt{s_{11} + 4Xs_{12} + 4X^2 s_{22}} \tag{5.10}$$

The t-test is then $(b_1 + 2b_2X)/s_b$ with $n - k - 1$ df, where $k = 2$ for X and X^2.

Table 5.2 provides an illustration of the analysis of simple slopes for the regression equation in Figure 5.3. Simple slopes are given at two standard deviations (-1.890) and one standard deviation (-0.945) below the mean of X, at the minimum of the curve $(X = -0.387)$, at $\bar{X} = 0.00$, and at one standard deviation $(X = 0.945)$ above the mean of X. At the mean of X and above, there is a strong positive relationship between X and Y that increases with increasing X; whereas at the point two standard deviations below X $(X = -1.890)$, there is a strong negative relationship.

The interpretation of an equation such as equation 5.1, which contains both linear and nonlinear functions of the same variable, is conceptually problematic—one cannot think of varying the second order term (X^2) while holding the linear term (X) constant. The simple slope approach handles this problem by repartitioning the predictable variation into a series of regressions of Y on X.[6]

Table 5.2
Simple Slopes at values of X for the Regression Equation $\hat{Y} = 4.993X + 6.454X^2 + 3.197$

a. Variance Covariance Matrix of Regression Coefficients

$$\mathbf{S}_b = \begin{matrix} & b_1 & b_2 \\ b_1 & 2.14449 & -0.10432 \\ b_2 & -0.10432 & 1.26567 \end{matrix}$$

b. Computation of Standard Error of Simple Slope at $X_H = 0.945$, where $s_X = 0.945$

(i) Simple slope: $b_1 + b_3X = 4.993 + 6.454(0.945) = 11.092$
(ii) Variance of simple slope: $s^2 = s_{11} + 4Xs_{12} + 4X^2s_{22}$
$= 2.144 + 4(0.945)(-0.104) + 4(.945)^2(1.266)$
$= 6.271$
(iii) Standard error of simple slope: $s_b = \sqrt{6.271} = 2.504$

c. t-Tests for Simple Slopes

X	Simple slope	Standard error	t
−1.890	−19.404	2.730	−7.11***
−0.945	−1.106	3.657	−0.41
−0.387	0.000	1.750	0.00
0.000	4.993	1.464	3.41***
0.945	11.092	2.504	4.43***

***$p < .001$;

The Progression of More Complex Curvilinear Equations Revisited

We now apply our prescriptions for post hoc probing to the progression of regression equations considered in the first section of this chapter. These equations, their simple slopes, and the simple slope variances are summarized in Table 5.1, Cases 2 though 4. Case 2 is a simple extension of Case 1. It is in Case 3 and Case 4 that we encounter the combination of higher order terms and crossproduct terms.

Case 2: Curvilinear X Relationship and Linear Z Relationship

Equation 5.2 represents an equation of this form and is reproduced here: $\hat{Y} = b_1X + b_2X^2 + b_3Z + b_0$. This equation, illustrated in Figure 5.2a, builds on the simple second order curvilinear equation just considered by the addition of the first order Z term. Differentiating with respect to X

yields $\partial \hat{Y}/\partial X = (b_1 + 2b_2X)$, the same simple slope for Y on X as in equation 5.1. This first (partial) derivative or simple slope does not contain Z, because Z does not interact with X. The only effect of changing the value of Z is to change the elevation of the whole curve, as can be seen in Figure 5.2a. Changing the elevation of the curve is reflected in changes in the intercept and in the value of $\hat{Y}$ at the maximum or minimum point of the curve. The value of X at which the minimum or maximum is reached is unaffected by Z.

The variance of the simple slope[7] is given in Table 5.1, Case 2. It is identical to that in the simple curvilinear regression equation (Case 1 of Table 5.1). The t-test of simple slopes at values of X follows directly with $n - k - 1$ df, where $k = 3$.

Case 3: Curvilinear X Relationship,
Linear Z Relationship,
and Linear by Linear XZ Interaction

This set of relationships is represented by equation 5.3, which is reproduced here: $\hat{Y} = b_1X + b_2X^2 + b_3Z + b_4XZ + b_0$.

This equation adds a linear by linear interaction between X and Z to the case we just considered. We first consider the regression of Y on X (Case 3a of Table 5.1); then we consider the regression of Y on Z for the same equation (Case 3b).

Case 3a: Reexpressed Regression Equation Showing Regression of Y on X. It is difficult conceptually to characterize the relationship of X and Z to Y from inspection of equation 5.3. Reordering and grouping terms in equation 5.3 presents the equation as a simple regression of Y on X at values of Z, a much more easily interpretable form. The rearranged equation has the same second order polynomial form (linear term, squared term, constant) as equation 5.1:

$$\hat{Y} = (b_1 + b_4Z)X + b_2X^2 + (b_3Z + b_0) \qquad (5.11)$$

The $(b_1 + b_4Z)$ coefficient in equation 5.11 takes on the same meaning in the simple regression equation as does b_1 in equation 5.3 and indicates the overall linear trend in the regression of Y on X at one value of Z. If $(b_1 + b_4Z)$ is positive, the simple regression has an overall upward linear trend; if it is negative, an overall downward linear trend. However, the nature of the curvature, measured by b_2, is independent of Z because Z does not interact with X^2.

We can use equation 5.11 to calculate the simple regression equations characterizing each of the three curves in Figure 5.2b(1). The overall regression equation is $\hat{Y} = 1.125X + 3.563X^2 + 3.608Z + 2.935XZ + 3.246$. With centered Z, and $s_Z = 2.200$, we can substitute values of Z into equation 5.11 to calculate the simple regression equations at Z_L, Z_M, and Z_H:

$$\text{In general, } \hat{Y} = (1.125 + 2.935Z)X + 3.563X^2 + 3.608Z + 3.246$$

$$\text{For } Z_L = -2.20, \quad \hat{Y} = [1.125 + (2.935)(-2.20)]X + 3.563X^2 + 3.608(-2.200) + 3.246$$

$$\hat{Y} = -5.332X + 3.563X^2 - 4.512$$

$$\text{For } Z_M = 0.00, \quad \hat{Y} = 1.125X + 3.563X^2 + 3.246$$

$$\text{For } Z_H = 2.20, \quad \hat{Y} = 7.582X + 3.563X^2 + 11.364$$

Because X is centered, the coefficient of X in each of the three simple regression equations can be interpreted as the conditional effect of X on Y at the mean of X for one value of Z, or the overall linear trend of the relationship of X to Y at one value of Z. From an inspection of Figure 5.2b(1), the overall linear trend is negative at Z_L, very slightly positive at Z_M, and more strongly positive at Z_H.

Case 3a: Simple Slopes. The reader should recall the distinction between coefficients found by reexpression of the overall regression equation, as in equation 5.11, and simple slopes. To determine the simple slope of the regression of Y on X at any single value of X, we differentiate equation 5.3 with respect to X:

$$\frac{\partial \hat{Y}}{\partial X} = b_1 + 2b_2X + b_4Z \tag{5.12}$$

The simple slope of Y on X at any single value of X now depends on the value of Z, as well as on the value of X. To interpret equation 5.3, we would calculate the simple slopes at all nine combinations of X_L, X_M, and X_H with Z_L, Z_M, and Z_H. To illustrate, consider the simple slope at $Z_H = 2.200$ and $X_L = -0.945$. Substituting into equation 5.12, we find that this simple slope is $1.125 + 2(3.563)(-0.945) + 2.935(2.200) =$

0.848. Table 5.3 provides a matrix summarizing the simple slopes at all nine combinations of X_L, X_M, and X_H crossed with Z_L, Z_M, and Z_H. Row 1, column 3 of the table shows the simple slope at X_L, Z_H to be 0.848 as we just calculated. The second row shows the regression of Y on X at the mean of centered X ($\bar{X} = 0$) for Z_L, Z_M, and Z_H. Note that these values are identical to the first order coefficients of the simple regression equations presented above. When $X = 0$, the regression coefficient $(b_1 + b_4Z)$ for X in reexpressed regression equation 5.11 and the simple slope of Y on X at a particular value of Z, that is, $(b_1 + 2b_2X + b_4Z)$ of equation 5.12, are equal.

When X is centered, both equations indicate the average regression of Y on X at the mean of X for particular values of Z, or the average slope of the regression of Y on X across all the cases in the sample.

The set of simple slopes presented in Table 5.3 leads to a useful summary of the outcome of the regression analysis. When X is low (row 1),

Table 5.3
Probing Simple Slopes in the Equation $\hat{Y} = 1.125X + 3.563X^2 + 3.61Z + 2.935XZ + 3.246$ (Simple Slopes Are Found Using the Expression $(b_1 + 2b_2X + b_4Z)$)

			(1) Z_L −2.200	(2) Z_M 0	(3) Z_H 2.200
		Simple slope	−12.065	−5.609	0.848
(1)	$X_L = -0.945$	standard error	2.628	2.819	3.786
		t	−4.591***	−1.990*	0.234
		Simple slope	−5.332	1.125	7.582
(2)	$X_M = 0$	standard error	2.324	1.532	2.153
		t	−2.294*	0.734	3.522***
		Simple slope	1.403	7.859	14.316
(3)	$X_H = 0.945$	standard error	3.900	2.839	2.502
		t	0.360	2.768	5.722***

$\mathbf{S}_b$: Covariance matrix of b coefficients:

$\mathbf{S}_b$ =	b_1	b_2	b_3	b_4
b_1	2.34649	0.01536	−0.41530	−0.08693
b_2	0.01536	1.58396	−0.04381	−0.49227
b_3	0.41530	−0.04381	0.43079	0.01174
b_4	−0.08693	−0.49227	0.01174	0.55213

$***p < .0001$; $**p < .01$; $*p < .05$

the regression of Y on X moves from highly negative to close to zero as Z increases. At the mean of X (row 2), the regression of Y on X passes from significantly negative to significantly positive as Z increases. When X is high (row 3), the regression of Y on X becomes increasingly positive as Z increases.

Table 5.3 illustrates the dramatic differences in the regression of Y on X as both X and Z vary. Once again, the use of simple slopes has led to a repartitioning of three sources of variation, those due to X, X^2, and XZ, leading to a straightforward interpretation of the regression analysis.

Case 3a: Standard Error and t-Test. The variance of the simple slope[8] is given in Table 5.1, Case 3a; its square root is the required standard error; and the t-test has $n - k - 1$ df, where $k = 4$.

Case 3a: Minimum or Maximum of Curve. Inspection of Figure 5.2b(1) suggests that the value of X at which $\hat{Y}$ has its minimum value for each regression curve depends on the value of Z. Setting equation 5.12 to zero and solving for X yields the following expression for the value of X at which $\hat{Y}$ is minimum for each of the regression curves:

$$X = \frac{-(b_1 + b_4 Z)}{2b_2} \tag{5.13}$$

Equation 5.13 produces the value of X corresponding to the minimum predicted value for U-shaped regression curves.[9] Substituting the value for $Z_L = -2.20$ into the equation, the minimum value of $\hat{Y}$ is found at $X = [(1.125 + (2.935)(-2.20)]/(-2)(3.563) = 0.75$. For Z_M and Z_H, the mimimum values of the regression curves are found at $X = -0.16$ and -1.06, respectively. In terms of our example, the value of self-concept at which self-disclosure (Y) began to increase occurs at ever lower levels of self-concept (X) as alcohol consumption (Z) increased from Z_L to Z_H.

Case 3a: Crossing Point. The curves corresponding to the simple regression equations for the regression of Y on X will all cross at a single point, $X = -b_3/b_4$. Substituting in the values of b_3 and b_4, we find that the simple regressions cross at the value $= -3.608/2.935 = -1.230$. In Figure 5.2b(1), X_L, the lower limit of X illustrated, is -0.95 (i.e., $\bar{X} = 0$ and $s_X = 0.95$). Hence the three curves intersect just below the value of X illustrated. If the range of X within which the three simple regressions of Y on X crossed were meaningful, then the interaction would be interpreted as disordinal. If, in contrast, the low value of X at which the curves cross is not meaningful, then the interaction would be interpreted as ordinal. In our example, if values below Z_L represented the self-concept scores of clinically depressed individuals (a different population), then we would state that the interaction was ordinal within the "normal" population above Z_L in self-concept. Note that if the b_4 term for the XZ interaction were zero, then the simple regressions would not cross. In this case, the equation reduces to the simpler regression equation 5.2, $\hat{Y} = b_1X + b_2X^2 + b_3Z + b_0$, which is depicted in Figure 5.2a(1).

Case 3b: The Regression of Y on Z. To this point we have considered only the regression of Y on X. As we have seen in previous chapters, probing the regression of Y on Z is also likely to be informative given the presence of the XZ interaction. The interaction signifies that the regression of Y on Z will also depend on values of X. For the present case of a simple XZ interaction, the regressions of Y on Z will all be linear. The simple regression lines of Y on Z at values of X are illustrated in Figure 5.2b(2).

Case 3b: Simple Slopes, Standard Error, and t-Test. The equation for the regression of Y on Z at levels of X is found by differentiating equation 5.3 with respect to Z:

$$\frac{\partial \hat{Y}}{\partial Z} = b_3 + b_4X \tag{5.14}$$

The variance of this simple slope[10] is given in Table 5.1, Case 3b. The standard error is the square root of this variance, and the t-test follows, with $n - k - 1$ df, where $k = 4$. The forms of both the simple slope and its variance, given in Table 5.1, Case 3b, are identical to those for the regression of Y on X at values of Z in equation 2.1: $\hat{Y} = b_1X + b_2Z + b_3XZ + b_0$ (see Table 4.4, Case 1b). This is so because in both equations 2.1 and 5.3 the relationship of Y to Z is linear both in the first order term and in the interaction. The more complex equation 5.3 differs from equation 2.1 only by the addition of the X^2 term, which does not enter into the regression of Y on Z.

Case 3b: Crossing Points. Figure 5.2b(2) shows that the simple regression lines do not cross at a single point, unlike our experience with previous simple linear by linear interactions. In equations containing second order terms in X, the crossing points of any two simple regression lines for the simple regressions of Y on Z depend upon specific values of X chosen. For equation 5.3, the value of Z at which two simple regression lines of Y on Z cross is

$$Z_{\text{cross}} = -[b_1 + b_2(X_i + X_j)]/b_4 \quad (5.15)$$

where X_i and X_j are the specific values of X chosen for examination.

Note again that if the b_4 coefficient for the XZ interaction is zero, the simple regression lines are parallel, as in Figure 5.2b(2). Further, if the b_2 coefficient for the X^2 term is zero, all simple regression lines will cross at the single value $-b_1/b_4$.

To illustrate the calculation of a crossing point, we substitute $X_H = 0.945$ for X_i and $X_M = 0$ for X_j into equation 5.15: $Z_{\text{cross}} = -[1.125 + 3.563(0.945 + 0)]/3.246 = -1.387$. This case is graphically depicted in Figure 5.2b(2), where the lowest value of Z plotted is $Z_L = -2.20$.

Case 4: Curvilinear X Relationship,
Linear Z Relationship, and
Curvilinear X by Linear Z Relationship

This set of relationships is represented by equation 5.4, which is reproduced here:

$$\hat{Y} = b_1X + b_2X^2 + b_3Z + b_4XZ + b_5X^2Z + b_0$$

Equation 5.4 adds an X^2Z term to equation 5.3.

Case 4a: Reexpressed Regression Equation to Show Regression of Y on X. As with equation 5.3, it is difficult at best to characterize the relationship of X and Z to Y from equation 5.4. To simplify interpretation, equation 5.4 may be rearranged into a simple regression equation showing the regression of Y on X at values of Z. In the rearranged equation we see that the regression coefficients for both X and X^2 vary as a function of the value of Z:

$$\hat{Y} = (b_1 + b_4Z)X + (b_2 + b_5Z)X^2 + (b_3Z + b_0) \quad (5.16)$$

As in Case 3a, in equation 5.16 the $(b_1 + b_4Z)$ coefficient provides the same information as the b_1 coefficient in the overall equation. The $(b_1 + b_4Z)$ coefficient represents the overall linear trend in the relationship of X to Y at a value of Z, paralleling its interpretation in equation 5.11. The

$(b_2 + b_5 Z)$ coefficient for X^2 in equation 5.16 conveys the same information as the b_2 coefficient in the overall equation. It represents the nature of the curvilinearity of the simple regression lines of Y on X at specific values of Z. If the value of $(b_2 + b_5 Z)$ is positive, the curve is concave upward; if negative, it is concave downward.

Figure 5.2c(1) illustrates regression equation $\hat{Y} = -2.042X + 3.000X^2 + 2.138Z + 2.793XZ + 1.960X^2Z + 3.502$. The figure is presented as three simple regressions, based on the rearranged form of equation 5.16:

$$\hat{Y} = (-2.042 + 2.793Z)X + (3.000 + 1.960Z)X^2 + (2.138Z + 3.502)$$

Substituting the values -2.200, 0, and 2.200 for Z_L, Z_M, and Z_H leads to values of both coefficients of equation 5.16 for each simple regression line. The value of $(b_1 + b_4 Z)$ varies from negative to positive as Z increases: -8.188 for Z_L, -2.042 for Z_M, and 4.104 for Z_H. This is consistent with the generally negative linear trend at Z_L, but generally positive linear trend at Z_H observed in Figure 5.2c(1). The values of $(b_2 + b_5 Z)$ are as follows: -1.313, 3.000, and 7.312, for Z_L, Z_M, and Z_H, respectively. At Z_L there is very slight downward curvature, but the curvature is upward for both Z_M and Z_H.

As a further aid to interpretation, both the $(b_1 + b_4 Z)$ coefficient and the $(b_2 + b_5 Z)$ coefficient may be tested for significance in each simple slope equation. Standard errors are computed according to the procedure given in Chapter 2 and employed throughout the text.[11] For example, for $(b_2 + b_5 Z)$, at Z_L, $t = -0.786$, ns; at Z_M, $t = 2.422$, $p < .05$; and at Z_H, $t = 4.853$, $p < .01$. These tests confirm the appearance of the regression of Y on X at Z_L as having a general upward linear trend that becomes increasingly concave upward as Z increases.

The reader should be aware that the linear coefficient $(b_1 + b_4 Z)$ and the curvilinear coefficient $(b_2 + b_5 Z)$ in equation 5.16 are not simple slopes. Rather these coefficients summarize the overall relationship of Y to X at particular values of Z. (In contrast simple slopes measure the regression of Y on X only at a single pair of X and Z values.)

For all the simple slopes presented in this chapter, the computer method developed in Chapter 2 may be used to find standard errors and simple slopes. We provide examples later in the chapter for the simple slopes presented. However, the computer method as presented in this text is only applicable to simple slopes and cannot be used to find the standard errors of the general linear and curvilinear coefficients in equation 5.16. Instead, the approach outlined in the optional section at the end of Chapter 2 and summarized in equation 2.10 must be used. The same is true for the general linear coefficient $(b_1 + b_4Z)$ in equation 5.11.

Case 4a: Simple Slopes, Standard Errors, and t-Tests. To find the simple slope of Y on X in equation 5.4, we compute the first (partial) derivative of equation 5.4 with respect to X:

$$\frac{\partial \hat{Y}}{\partial X} = b_1 + 2b_2X + b_4Z + 2b_5XZ \tag{5.17}$$

The simple slope of Y on X depends on the values of both X and Z. The variance of the simple slope[12] is given in Table 5.1, Case 4a. The t-test follows with $n - k - 1$ df, where $k = 5$.

The simple slopes corresponding to all possible combinations of X_L, X_M, and X_H with Z_L, Z_M, and Z_H are calculated by substituting the appropriate X and Z values into equation 5.17. The matrix containing the nine simple slopes for our numerical example is presented in Table 5.4. For example, for the combination X_L, Z_M (row 1, column 2 of Table 5.4), the simple slope is -7.711. For the combination X_M, Z_M at the means of both X and Z (row 2, column 2), the simple slope is -2.042, the same value as the b_1 coefficient for X in the overall equation. The standard errors, and the corresponding t-tests for each of the nine combinations of values of X and Z, are also provided and should be compared with Figure 5.2c(1). For X_L, we observe that the simple slope of the regression curve becomes increasingly negative as the value of Z increases. In contrast, for X_M, the simple slope of the regression curve has a negative value at Z_L, does not differ from 0 at Z_M, and becomes increasingly positive thereafter as Z increases. Finally, for X_H, the simple slope of the regression curve is highly negative at Z_L and rapidly changes to become positive at Z_M and highly positive at Z_H.

Case 4a. Maximum and Minimum of Curves. To find the maximum or minimum of each simple regression curve, we set simple slope equation 5.17 equal to 0 and solve for X.

$$b_1 + 2b_2X + b_4Z + 2b_5XZ = 0$$

so that

$$X = \frac{-(b_1 + b_4Z)}{2(b_2 + b_5Z)} \tag{5.18}$$

Table 5.4
Probing Simple Slopes in the Equation $\hat{Y} = -2.042X + 3.000X^2 + 2.138Z + 2.793XZ + 1.960X^2Z + 3.502$ [Simple Slopes Found Using the Expression $(b_1 + 2b_2X + b_4Z + 2b_5XZ)$]

			(1) Z_L −2.200	(2) Z_M 0	(3) Z_H 2.200
(1)	$X_L = -0.945$	Simple slope	−5.706	−7.711	−9.716
		standard error	2.963	2.801	4.439
		t	−1.925*	−2.753**	−2.189*
(2)	$X_M = 0$	Simple slope	−8.188	−2.042	4.104
		standard error	2.368	1.669	2.256
		t	−3.457***	−1.224	1.819
(3)	$X_H = 0.945$	Simple slope	−10.669	3.627	17.924
		standard error	4.731	2.946	2.586
		t	−2.255*	1.231	6.930***

$\mathbf{S}_b$: Covariance matrix of b coefficients:

	b_1	b_2	b_3	b_4	b_5
b_1	2.78447	0.11045	−0.14763	−0.05919	−0.33314
b_2	0.11045	1.53329	0.00252	−0.46696	−0.05926
b_3	−0.14763	0.00252	0.52842	0.02239	−0.15468
b_4	−0.05919	−0.46696	0.02239	0.52960	−0.01487
b_5	−0.33314	−0.05926	−0.15468	−0.01487	0.20618

***$p < .001$; **$p < .01$; *$p < .05$

To illustrate, for the regression equation $\hat{Y} = -2.042X + 3.000X^2 + 2.138Z + 2.793XZ + 1.960X^2Z + 3.502$, we substitute into equation 5.18 to calculate the minimum at $Z_L = -2.200$.

$$X = \frac{-[(-2.042) + (2.793)(-2.200)]}{2[3.000 + (1.960)(-2.200)]} = -3.120 \text{ for } Z_L$$

Inspection of Figure 5.2c(1) indicates that this point corresponds to the *maximum* predicted value of Y. Similar calculations yield $X = 0.34$ for Z_M and $X = -0.281$ for Z_H, both of which, from inspection of Figure 5.2c(1), correspond to the *minimum* predicted values of their respective regression curves.[13] As can be seen from inspection of Figure 5.2c(1), these results, coupled with the examination of the simple regression curves of Figure 5.2c(1) and their simple slopes, suggest that different relationships of self-concept to self-disclosure exist at low versus moderate and higher levels of alcohol consumption.

Case 4a: Crossing Point. We use the usual strategy to determine the value of X at which the curves for the regression of Y on X at values of Z cross. Two values of Z, Z_i and Z_j, are chosen and substituted into the regression equation 5.4; then the two equations are set equal. We find that there are now two possible crossing points, represented by the following equations:

$$X_{\text{cross}(1)} = \left[-b_4 + \left(b_4^2 - 4b_3b_5\right)^{1/2}\right]/2b_5 \qquad (5.19)$$

$$X_{\text{cross}(2)} = \left[-b_4 - \left(b_4^2 - 4b_3b_5\right)^{1/2}\right]/2b_5 \qquad (5.20)$$

When the b_5 coefficient for the X^2Z term is zero, the curves will not cross. However, even when b_5 is nonzero, there may still be no point at which the two regression curves cross. In fact, this is true for the present case, as is illustrated in Figure 5.2c(1). Evaluation of equation 5.19 for this case does not produce a real number for a solution; the solution is an imaginary number:

$$\begin{aligned} X_{\text{cross}(1)} &= -\left[2.793 + \left[2.793^2 - 4(2.138)(1.960)\right]^{1/2}\right]/2(1.960) \\ &= \left[-2.793 + (-8.961)^{1/2}\right]/3.92 \end{aligned}$$

Evaluation of equation 5.20 also fails to produce a real number solution. Thus the regression curves do not cross.

On p. 82, the crossing point for the regression of Y on X in Case 3a was given as $X = -b_3/b_4$. The crossing points in equations 5.19 and 5.20 for Case 4a do not appear to be straightforward generalizations of that for Case 3a. However, the discontinuity is only apparent. The limit of equation 5.19 and equation 5.20 as b_5 approaches zero is, in fact, $\pm b_3/b_4$.

Case 4b: The Regression of Y on Z. Given the significant interactions involving Z, probing the regression of Y on Z at levels of X is also likely to be informative. The formulas for the simple slope of this regression line and its variance are given in Table 5.1, Case 4b. As can be seen in Figure 5.2c(2), the regressions of Y on Z at levels of X are linear in form.

Each pair of simple regression lines cross at values of Z that depend upon the values of X in question:

$$Z_{\text{cross}} = \frac{-[b_1 + b_2(X_i + X_j)]}{b_4 + b_5\,(X_i + X_j)} \tag{5.21}$$

For example, when $X_i = X_H = 0.945$ and $X_j = X_M = 0$,

$$Z_{\text{cross}} = \frac{-[(-2.042) + 3.000(0.945 + 0)]}{[2.793 + 1.960(0.945 + 0)]} = -0.17$$

Coefficients of Simple Slopes by Computer

The simple slopes and the corresponding standard errors and t-tests for all the analyses in Tables 5.2, 5.3, and 5.4 may be calculated by computer. The approach directly extends the three step procedure presented in Chapter 2. We will consider the probing of equation 5.4, which in our example includes a significant X^2Z interaction. The outcome of the simple slope analysis for the numerical example is presented in Table 5.4; the parallel computer analysis is presented in Table 5.5.

The significant X^2Z interaction implies that each regression of Y on X depends on the specific values of both X and Z. Consequently, we need to specify these values and we will use all combinations of X_L, X_M, and X_H with Z_L, Z_M, and Z_H as before. Recall that X and Z are centered; $X_M = 0$ and $Z_M = 0$ so that new transformed variables are not needed for these values.

1. Our first step is to transform the original X and Z variables so they are evaluated at the conditional values of interest. The transformed variables are created from X and Z by subtracting conditional values CV_X and CV_Z, respectively. In this case we have the following:

(a) XABOVE $= X - (0.945)$ for the regression of Y on X at $CV_X = 0.945$, one standard deviation above the mean of X;
(b) XBELOW $= X - (-0.945)$ for the regression of Y on X at $CV_X = -0.945$, one standard deviation below the mean of X;
(c) ZABOVE $= Z - (2.200)$ for the regression of Y on X at $CV_Z = 2.200$, one standard deviation above the mean of Z; and
(d) ZBELOW $= Z - (-2.200)$ for the regression of Y on X at $CV_Z = -2.200$, one standard deviation below the mean of Z.

Table 5.5
Computation of Simple Slope Analysis by Computer for the X^2Z Interaction in the Regression Equation $\hat{Y} = b_1X + b_2X^2 + b_3Z + b_4XZ + b_5X^2Z + b_0$

---

a. Overall Analysis with Centered X and Centered Z

(i) Means and standard deviations

	Mean	Std Dev
Y	8.944	29.101
X	0.000	0.945
X²	0.890	1.230
Z	0.000	2.200
XZ	0.861	2.086
X²Z	0.187	3.944

(ii) Regression analysis

Variable	B	SE B	T	Sig T
X	−2.041992	1.668673	−1.224	.2218
X²	2.999519	1.238259	2.422	.0159
Z	2.138031	0.726923	2.941	.0035
XZ	2.793482	0.727738	3.839	.0001
X²Z	1.960267	0.454067	4.317	.0000
(Constant)	3.501767	1.586818	2.207	.0279

b. Computation of XABOVE, XBELOW, ZABOVE, ZBELOW, and Crossproduct Terms Required for Simple Slope Analysis

```
COMPUTE XABOVE=X-(.945)
COMPUTE XBELOW=X-(-.945)
COMPUTE ZABOVE=Z-(2.20)
COMPUTE ZBELOW=Z-(-2.20)
COMPUTE X2A=XABOVE*XABOVE
COMPUTE X2B=XBELOW*XBELOW
COMPUTE XZA=X*ZABOVE
COMPUTE XZB=X*ZBELOW
COMPUTE XAZ=XABOVE*Z
COMPUTE XBZ=XBELOW*Z
COMPUTE XAZA=XABOVE*ZABOVE
COMPUTE XAZB=XABOVE*ZBELOW
COMPUTE XBZA=XBELOW*ZABOVE
COMPUTE XBZB=XBELOW*ZBELOW
COMPUTE X2ZA=X2*ZABOVE
COMPUTE X2ZB=X2*ZBELOW
COMPUTE X2AZA=X2A*ZABOVE
COMPUTE X2AZB=X2A*ZBELOW
COMPUTE X2BZA=X2B*ZABOVE
COMPUTE X2BZB=X2B*ZBELOW
```

Table 5.5, continued

c. Regression Analysis with XABOVE and ZBELOW, Yielding Simple Slope Analysis at X_H and Z_L (regression of Y on X one standard above the mean of X and one standard deviation below the mean of Z)

(i) Means and standard deviations

	Mean	Std Dev
Y	8.944	29.101
XABOVE	−0.945	0.945
X2A	1.783	2.103
ZBELOW	2.200	2.200
XAZB	−1.218	3.139
X2AZB	2.484	5.678

N of Cases = 400

(ii) Regression analysis

Variable	B	SE B	T	Sig T
XABOVE	−10.669354	4.730779	−2.255	0.0247
X2A	−1.313069	1.670902	−0.786	0.4324
ZBELOW	6.528441	0.952239	6.856	0.0000
XAZB	6.498388	1.099950	5.908	0.0000
X2AZB	1.960268	0.454067	4.317	0.0000
(Constant)	−10.111838	3.627950	−2.787	0.0056

d. Regression Analysis with X and ZABOVE, Yielding Simple Slope Analysis at $\overline{X}$ and Z_H (Regression of Y on X at the Mean of X and One Standard Deviation Above the Mean of Z)

(i) Means and standard deviations

	Mean	Std Dev
Y	8.944	29.101
X	.000	.945
X2	.890	1.230
ZABOVE	−2.200	2.200
XZA	.861	2.801
X2ZA	−1.771	4.403

(ii) Regression analysis

Variable	B	SE B	T	Sig T
X	4.103670	2.255509	1.819	0.0696
X2	7.312106	1.506802	4.853	0.0000
ZABOVE	2.138032	.726923	2.941	0.0035
XZA	2.793482	.727738	3.839	0.0001
X2ZA	1.960267	.454067	4.317	0.0000
(Constant)	8.205437	2.260164	3.630	0.0003

2. Crossproducts of transformed variables with themselves (for squared terms), with each other, and with original variables X and Z are formed. The required crossproducts are those corresponding to each of the terms in the simple slope equation under consideration. For example, for the regression of Y on X one standard deviation above the mean of X (X_H) and one standard deviation below the mean of Z (Z_L), the following crossproducts are required:

X2A = XABOVE * XABOVE

XAZB = XABOVE * ZBELOW

X2AZB = XABOVE * XABOVE * ZBELOW = X2A * ZB

3. The regression analysis is performed using the transformed variables and their crossproducts. In each case the b_1 coefficient for the X predictor represents the regression of Y on X at the conditional values of X and Z that have been specified.

Table 5.5 illustrates the use of the SPSS-X regression program to test the simple slopes at two of the nine combinations of X_L, X_M, and X_H with Z_L, Z_M, and Z_H. Table 5.5a shows the overall regression analysis with the centered predictors. Note that in the centered case, the test of b_1 corresponds to the test of the simple slope for the regression of Y on X evaluated at the values X_M and Z_M. Table 5.5b shows computer code that will generate all needed terms required to perform the nine tests of simple slopes presented in Table 5.4. Table 5.5c presents the test of the simple slope of the regression of Y on X at X_H and Z_L, corresponding to the results of the parallel matrix-based test in Table 5.4 (row 3, column 1 of the matrix of simple slopes). Finally, Table 5.5d shows the regression of Y on X at the mean of X (X_M) and Z_H and corresponds to the test of simple slopes in Table 5.4 (row 2, column 3). The identical results of the matrix-based and computer-based analyses illustrate the equivalence of the two procedures.

Three Final Issues

Curvilinearity Versus Interaction

Darlington (1990) has clearly pointed out the difficulty of distinguishing between regression equations including an interaction term, $\hat{Y} = b_1X + b_2Z + b_3XZ + b_0$, and regression equations including a curvilinear effect, $\hat{Y} = b_1X + b_2X^2 + b_0$, in data sets in which X and Z are highly correlated. In such instances the X^2 and XZ terms will be highly correlated. Thus it will often be difficult and sometimes impossible to distinguish between the two regression models on statistical grounds, even in large samples. Lubinski and Humphreys (1990) provide an illustration of this problem in attempting to distinguish between a curvilinear model and an interaction model for predicting scores on an advanced mathematics test for a large national sample of high school students.

Within a given sample, there is little researchers may do to distinguish between the two interpretations. They may argue that the model supported by the better articulated substantive theory should be retained. Or, they may argue that the second model, $Y = b_1X + b_2X^2 + b_0$, involves only one variable (X) and has one fewer parameter and should therefore be accepted on the grounds of parsimony. A better solution is to locate a new sample in which X and Z are less correlated or to sample cases from the population in a manner designed to reduce the correlation between X and Z. When X and Z contribute nonredundant information, then models representing interaction versus curvilinearity can be distinguished.

What Terms Should Be Included in the Regression Equation?

The example used throughout this chapter was based on a simulation, so that the actual equation that generated the data is known. Typically, however, the researcher must decide which terms to include in the regression equation. This decision should be guided by prior theory and empirical research in the substantive area. In addition, the analyst may also have new hypotheses about potential effects that should be included in the equation. Nonetheless, in the absence of definitive knowledge about the area, the researcher can easily include too few or too many terms in the equation relative to the "true" regression equation. Each of these outcomes has different benefits and consequences.

Omitting higher order terms whose true effects are nonzero from the equation biases the lower order coefficients. Each lower order term (e.g., b_1X) that is tested includes unique variance that is due to the lower order term plus all variance that is shared with the omitted nonzero higher order terms. This problem can be illustrated by comparing two regression equations using our simulated data set. If we estimate an equation containing only X and Z terms, we find

$$\hat{Y} = 1.923X + 3.726Z + 8.944$$

In contrast, if we estimate equation 5.4, which generated the data, we find

$$\hat{Y} = -2.042X + 2.138Z + 2.996X^2 + 2.793XZ + 1.960X^2Z + 3.502$$

We note the dramatic change in the coefficients for the X and Z terms and the intercept, illustrating the bias that is introduced by omitting these nonzero higher order terms from the equation. Neter, Wasserman, and Kutner (1989) discuss several methods of plotting residuals that are useful in detecting this problem.

The inclusion of higher order terms whose true value is zero should not bias the estimates of lower order terms. In Chapter 3 we pointed out that when X and Z are centered and bivariate normally distributed, then the correlations of X with X^2, Z with Z^2, and X and Z with XZ are zero. It would seem then that under these conditions including a higher order terms, say X^2 in equation 5.2, or XZ in equation 2.1, should have essentially no effect on the estimates of the X and Z effects. However, with higher order terms, if the first order predictors are even moderately correlated, then first and third order terms (e.g., Z with X^2Z) will be highly correlated, even for centered predictors. The same is true for second and fourth order[14] terms (e.g., X^2 with XZ; X^2 with X^2Z^2). These interpredictor correlations will introduce instability into regression coefficients; the correlations exist even after variables are centered (Dunlap & Kemery, 1987; Marquardt, 1980). To illustrate, if we estimate equation 5.5, which contains all the terms from 5.4 plus three terms known to be zero in our simulated data set (i.e., Z^2, XZ^2, X^2Z^2), we find

$$\hat{Y} = -1.027X + 2.412Z + 3.497X^2 + 0.411Z^2 + 1.929XZ - 0.534XZ^2 + 2.725X^2Z - 0.003X^2Z^2 + 1.795$$

We note that (a) the coefficients for each of the five nonzero terms has changed somewhat from the values estimated above for equation 5.4 (though all significance levels remain highly similar), and (b) one of the zero terms in the population is significant in the sample (for XZ^2, $p < .05$). The source of the significance is revealed in the pattern of intercorrelations of XZ^2 with other variables. The XZ^2 term has a low positive zero order correlation with the criterion ($r = 0.246$) but a very high zero order correlation with the other third order term X^2Z ($r = 0.817$). The significant *negative* coefficient for XZ^2 is due to an unexpected suppressor effect.[15]

We thus caution researchers against routinely including higher order terms not expected from theory on two grounds. As we have just seen, such terms may introduce high interpredictor correlations that cause instability of regression coefficients and anomalous results, even with centered predictors. In addition, the inclusion of extraneous terms lowers the statistical power of the tests of all terms in the equation. Chapter 6 discusses the procedures and the conditions under which nonsignificant higher order terms can be dropped from regression equations to insure higher power tests of the remaining terms.

Other Methods of Representing Curvilinearity

In this book we have emphasized the use of regression analysis as a means of testing models developed from strong theory in which the equations take the form of polynomial expressions. These equations represent the forms typically hypothesized by current theory in most areas of the social sciences. At the same time, other forms of nonlinear interactions are occasionally hypothesized and regression analyses may be conducted in a more exploratory vein. Below, we briefly note some of the approaches that may be taken to other forms of theoretically predicted nonlinear interactions and to the exploratory use of regression analysis in practical prediction problems.

Other Forms of Theory-Based Nonlinear Interactions

The simplest approach to nonlinear regression problems, where possible, is to test models that are linear in the parameters or in which the equation has been linearized. For example, suppose theory predicted that $X^* = \log(X)$ and $Z^* = (Z)^{1/2}$. Furthermore, the theory predicts that the transformed variables interact to predict Y:

$$\hat{Y} = b_1X^* + b_2Z^* + b_3X^*Z^* + b_0$$

This equation follows the standard form we have estimated throughout the book. Note that an alternative representation of the equation is as follows:

$$\hat{Y} = b_0 + b_1 \log(X) + b_2(Z)^{1/2} + b_3 \log(X)(Z)^{1/2}$$

This second equation is linear in the parameters and would be estimated using ordinary least squares (OLS) regression and can be interpreted in the transformed [e.g., log (X)], but not the original X scaling in terms of the prescriptions we have outlined in this book.

An example of a linearized equation is provided by Wonnacott and Wonnacott (1979) who note that economic theory describes the Cobb–Douglas production function as follows: $Q = b_0 K^{b_1} L^{b_2} u$, where Q is the quantity produced; K is capital; L is labor; b_0, b_1, and b_2 are the nonlinear regression parameters to be estimated; and u is the multiplicative error term. If we take the logarithms of both sides of the equation, we discover

$$\log(Q) = \log(b_0) + b_1 \log(K) + b_2 \log(L) + \log(u)$$

which is equivalent to the easily estimated linear regression equation:

$$Y = b_0^* + b_1 X + b_2 Z + e$$

The regression coefficients can be estimated through OLS regression and the value of b_0 can be calculated from b_0^* (by taking the antilog).

More complex nonlinear regression equations that cannot be linearized can also be estimated using iterative, numerical search procedures. These procedures are beyond the scope of the present book. Introductions to the estimation, testing, and interpretation of such equations can be found in Judge, Hill, Griffiths, Lutkepul, and Lee (1982), Kmenta (1986), and Neter et al. (1989), while Gallant (1987) provides a more advanced treatment. Finally, note that in each of the procedures outlined above, the researcher's interest is in developing unbiased or minimally biased estimates and tests of the regression coefficients of a theoretically specified nonlinear model. Such tests are conducted to provide evidence for or against the viability of theory-based hypotheses.

Exploratory Regression Analyses in Practical Prediction

Researchers interested in practical prediction problems often encounter data that are not amenable to simple linear regression. These researchers have developed a variety of graphical and statistical techniques for simplifying regression equations through nonlinear transformations of the variables (e.g., Atkinson, 1985; Box & Cox, 1964; Daniel & Wood, 1980). These techniques minimize problems in the data (e.g., outliers) and maximize R^2. They typically do not provide unbiased tests of the regression coefficients in models based on substantive theory. Indeed, data transformations in the interest of simplifying the regression equation can often eliminate interesting curvilinear or interactive effects predicted on the basis of theory. Thus researchers need to keep in mind the nature of their work, theory testing or practical prediction, when they are choosing a strategy for nonlinear regression analysis.

Summary

For complex regression equations involving significant higher order terms and their interactions, interpretation is facilitated by repartitioning the variance of the original regression equation. A three-stage process has been employed for recasting regression equations. First, the overall regression equation is rearranged to show the regression of the criterion on one of the predictors, with the regression coefficients being expressed as linear combinations of the coefficients of the overall regression equation. Equations 5.11 and 5.16 represent examples of such reexpressions into X and X^2 components for the regression of Y on X. Second, values of the other predictors are then substituted into the reorganized expressions to generate a series of simple regression equations, such as those illustrated in Figure 5.2. Finally, simple slopes for the regression of Y onto a single predictor at specified values of that and/or other predictors are computed, for example, the regression of Y on X at values of X and Z in equations 5.3 and 5.4. These simple slopes provide insights into the trends represented by each simple regression equation.

Notes

1. In very rare instances, based on strong theory, only the quadratic component might be included. To illustrate, suppose we had a model for judgments of size that indicated that size was judged on the basis of area, and not on the basis of height or width of objects alone. Then in a regression equation predicting size judgments from area, one might wish to omit the first order terms of height and width. The first order terms might be omitted if both strong theory plus prior empirical evidence indicated that individuals do not rely at all on linear extent in the judgment of size (we thank David Kenny for this example).

2. If there are nonlinear relationships between predictors and the criterion, and these relationships are not reflected in the regression equation, they may be detected with regression diagnostics applied to residuals (see, for example, Belsley, Kuh, & Welsh, 1980; Bollen & Jackman, 1990; Cook & Weisberg, 1980; Stevens, 1984), and the regression equation appropriately respecified.

3. Briefly, the operation of differentiation is applied separately to each term in the regression equation. For polynomial expressions like the ones we are considering, the first (partial) derivative (with respect to X) of the general term $b_i X^n Z^m$ is $(b_i)(nX^{n-1})(Z^m)$. Standard calculus texts (e.g., Thomas, 1972) provide detailed discussions of this operation for readers interested in pursuing this topic in more depth.

4. Readers familiar with calculus can confirm whether the value of X corresponds to the maximum or minimum value of $\hat{Y}$ by computing the value of the second derivative of the original curvilinear regression equation, $\hat{Y} = b_1 X + b_2 X^2 + b_0$. This quantity is equal to $2b_2$ and indicates that $\hat{Y}$ will be a minimum when b_2 is positive, whereas $\hat{Y}$ will be a maximum when b_2 is negative.

5. Our general matrix-based procedure for testing simple slopes applies here. The expression $(b_1 + 2b_2 X)$ is a linear combination of regression coefficients. Thus an estimate of its standard error is provided by the square root of the expression $s_b^2 = \mathbf{w}' \mathbf{S}_b \mathbf{w}$. For the simple curvilinear regression equation 5.1, the weight vector $\mathbf{w}'$ for b_1 and b_2 is $[1 \quad 2X]$, and $\mathbf{S}_b$ is the 2×2 covariance matrix of the regression coefficients.

6. Stimson, Carmines, and Zeller (1978) have provided a method of recasting polynomial regression equations to render them more interpretable. The simple polynomial regression $\hat{Y} = b_1 X + b_2 X^2 + b_0$ is rewritten as follows:

$$\hat{Y} = M + b_2(F - X)^2$$

where

$$M = \text{minimum (or maximum) value of the criterion,}$$

$$= b_0 - \frac{b_1^2}{4b_2^2}$$

$$F = \text{value of } X \text{ where minimum (or maximum) performance is produced,}$$

$$= \frac{-b_1}{2b_2}$$

For example, if one considers the U-shaped relationship between self-concept (X) and self-disclosure (Y), the rewritten polynomial equation would indicate the minimum value of self-disclosure (M) at the value of self-concept (F) at which that minimum M was achieved.

7. For the general matrix-based approach to the variance of the simple slope of Y on X in Case 2, $\mathbf{w}' = [1 \quad 2X \quad 0]$ and $\mathbf{S}_b$ is 3×3.

8. For the general matrix-based approach to the variance of the simple slope of Y on X in Case 3a, $\mathbf{w}' = [1 \quad 2X \quad 0 \quad Z]$ and $\mathbf{S}_b$ is 4×4.

9. The second (partial) derivative of equation 5.3 with respect to X is b_2. If this second derivative is positive, then a minimum has been identified; if it is negative, a maximum has been identified (see note 4 above).

10. For the general matrix-based approach to the variance of the simple slope of Y on Z in Case 3b, $\mathbf{w}' = [0 \quad 0 \quad 1 \quad X]$ and $\mathbf{S}_b$ is the same 4×4 matrix as for Case 3a.

11. The variance of the coefficient $(b_1 + b_4 Z)$ is calculated using the general matrix approach, where $\mathbf{w}' = [1 \quad 0 \quad 0 \quad Z \quad 0]$ and $\mathbf{S}_b$ is 5×5. The t-test has $n - k - 1$ df, where $k = 5$. For $(b_2 + b_5 Z)$, $\mathbf{w}' = [0 \quad 1 \quad 0 \quad 0 \quad Z]$, and $\mathbf{S}_b$ is the same 5×5 matrix; t has the same df.

12. For the general matrix-based approach to the variance of the simple slope of Y on X in Case 4a, $\mathbf{w}' = [1 \quad 2X \quad 0 \quad Z \quad 2XZ]$ and $\mathbf{S}_b$ is 5×5.

13. The second partial derivative of equation 5.4 with respect to X is

$$\frac{\partial^2 \hat{Y}}{\partial X^2} = 2(b_2 + b_5 Z)$$

This equation indicates that the direction of curvature depends on the value of Z. The value of X obtained from equation 5.18 will be a minimum when $b_2 + b_5 Z > 0$ and a maximum when $b_2 + b_5 Z < 0$. At $Z_L = -2.20$, the second derivative is $2[3.000 + (1.96)(-2.20)] = -1.312$, indicating a maximum; hence the regression of Y on X is slightly concave downward. For Z_H, the second derivative is 15.75; the regression of Y on X is concave upward.

14. For multivariate normal variables X, Z, and W, all odd moments vanish (e.g., XZW, X^2Z, X^2Z^2W), whether or not X, Z, and W are intercorrelated. If X, Z, and W are intercorrelated, then even moments do not vanish (e.g., X^2Z^2, $X^2Z^2W^2$). See the appendix of Kenny and Judd (1984) for a summary of moments of multivariate normal distributions.

15. A suppressor variable is one that is generally uncorrelated with the criterion, is highly correlated with another predictor, and increases the predictability of the other predictor by its inclusion in the equation. The suppressor typically has a significant negative regression coefficient.

6 Model and Effect Testing with Higher Order Terms

The inclusion of nonlinear and interactive terms produces complex regression equations like those considered in Chapters 4 and 5. The discussion thus far has focused on the probing and interpretation of the highest order term in the full model in which all terms are included. In this chapter we address in more detail the issues involved in simplifying regression equations by dropping nonsignificant higher order terms from the model. We also consider several global tests designed to compare the full model with reduced (simpler) models that may provide an equally good fit of the data. Finally, we consider sequential testing strategies aimed at simplifying complex regression equations containing higher order interactions. Throughout this chapter it is important for the reader to keep in mind why the specific term(s) being probed were included in the regression equation. Different analytic strategies and different intepretations may be appropriate depending on whether the source of the term was strong theory or data exploration.

Some Issues in Testing Lower Order Effects in Models Containing Higher Order Terms

A central issue in testing complex regression equations is that the lower order effects and the interactions are not typically independent. The variance that is shared by terms in the equation could potentially be apportioned to the higher and lower order effects in several different ways. This issue has stimulated a sizeable literature comparing strategies for testing overlapping effects in MR (see, e.g., Allison, 1977; Cleary & Kessler, 1982; Cohen & Cohen, 1983; Darlington, 1990; Lane, 1981; Pedhazur, 1982; Piexoto, 1987). Interestingly, a large parallel literature in ANOVA discusses strategies for partitioning the variance and testing the effects in factorial designs with unequal (nonproportional) cell sizes (see, e.g., Appelbaum & Cramer, 1974; Cramer & Appelbaum, 1980; Herr & Gaebelein, 1978; Overall, Lee, & Hornick, 1981; Overall & Spiegel, 1969; Overall, Spiegel, & Cohen, 1975). These designs have the same problem that is typical of complex multiple regression equations: The partitioning of variance is not unambiguous.

To illustrate this issue and the questions it engenders, we will consider the simplest case, represented by the now familiar two variable regression equation containing an interaction. This equation is reproduced as equation 6.1 below.

$$\hat{Y} = b_1X + b_2Z + b_3XZ + b_0 \tag{6.1}$$

With this equation as our starting point, a number of simpler models can also be generated, presented here as equations 6.2, 6.3, and 6.4.

$$\hat{Y} = b_1X + b_2Z + b_0 \tag{6.2}$$

$$\hat{Y} = b_1X \qquad\qquad + b_0 \tag{6.3}$$

$$\hat{Y} = \qquad\qquad b_2Z + b_0 \tag{6.4}$$

Note that the b_1X terms in equations 6.1, 6.2, and 6.3 will not generally be equal. In equation 6.3, b_1 reflects all variance shared between X and Y. In equation 6.2, b_1 reflects all variance shared between X and Y over and above the variance shared between Z and Y. Finally, in equation 6.1, b_1 reflects the unique variance shared between X and Y after the effects of Z and XZ on Y have been removed.[1] Thus the b_1s in the three equations have different meanings and *may* vary substantially in magnitude, depending on interpredictor correlation and the distribution of the variables.

Thus far in the book we have discussed the interpretation of only the b_1 and b_2 coefficients of equation 6.1. Each lower order effect was interpreted assuming the presence of the interaction in the equation.[2] Testing the significance of each term in equation 6.1 provides a test of the unique variance attributable to each of the effects. This strategy is always the method of choice if b_3 is significant. The central question in this case is what interpretation, if any, should be given to the lower order terms. This first question, which we have already considered and whose answer we will briefly review below, has been the focus of an extensive debate in the literature (Appelbaum & Cramer, 1974; Cohen, 1978; Darlington, 1990; Finney, Mitchell, Cronkite, & Moos, 1984; Overall et al., 1981; Pedhazur, 1982).

Continuing with our example, suppose that b_3 is not significant. This outcome leads to a second question that has also been the focus of considerable debate in the literature (e.g., Cramer & Appelbaum, 1980; Finney et al., 1984; Overall et al., 1981): Should the XZ term be dropped from the equation and the b_1 and b_2 coefficients be estimated using reduced equation 6.2? In answering this question, it will be useful to distinguish between two different cases: (a) Strong theory makes predictions only about the X and Z effects, but the analyst included the XZ term to explore what would happen. (b) Strong theory makes predictions about the XZ effect that were not supported in the present study.

Question 1. Interpretation of Lower Order Terms when b_3 Is Significant

The interpretation of lower order terms in the presence of an interaction has been previously discussed in Chapters 3 and 5. To review briefly, the first order effects do *not* represent a *constant effect* across the range of the second variable. Rather, for centered variables, they represent the *conditional* effect of the variable at the mean of the second variable. They may also be interpreted as the average effect defined as the mean of the simple slopes evaluated for each case in the sample (Darlington, 1990; Finney et al., 1984). Thus, so long as the predictor variables are centered, the lower order effects have a clear interpretation.

There are many situations in which the average value of the regression of Y on X at the mean value of Z (i.e., the *conditional* effect of X at $\bar{Z}$) in the sample will be useful. For example, if poor school children receive or do not receive a hot lunch program (X) regardless of the quality of their home diet (Z), it is useful to know whether, on average in the sample at least, there is a salutory effect of the hot lunch program on some health outcome (Y). Stated in terms of the interpretation of b_1 as a conditional effect at $Z = 0$ for centered Z, it is useful to know whether there is a salutory effect of the hot lunch program for children with an average (typical) home diet. Consequently, we recommend that variables be centered and that conditional effects be tested. At the same time, we believe that authors have an obligation to the reader to explain the meaning of conditional effects and how they differ from more familiar constant main effects. Restricting the presentation of the tests of the conditional effects to the context of post hoc probing of the interaction would help minimize the possibility of misinterpretion of such effects.

Question 2. Should Lower Order Coefficients Be Tested in Reduced Models when b_3 Is Nonsignificant?

The second question concerns the proper procedure to follow when the test of the b_3 coefficient for the interaction is nonsignificant. This question centers on whether or not the interaction should be eliminated from the equation, with the first order X and Z terms now being tested using equation 6.2. To begin to answer this question, we need to consider the tradeoff between two desirable properties of statistical estimators: unbiasedness and efficiency.

Unbiasedness means that the estimators, in this case the sample regression coefficients, will on the average equal the population value of the corresponding parameters. In standard regression models, the major source of bias is the omission of terms from the regression equation that represent true effects in the population (specification error). High efficiency means that the size of the standard error of the estimator relative to other estimates of the same standard error will be small. Including additional terms in a regression equation that in fact have no relationship to the criterion has no effect on bias: All of the regression coefficients will continue to be unbiased estimators of their respective population values. However, the introduction into the equation of unnecessary terms having no relationship to the criterion in the population has the result of lowering, sometimes appreciably, the efficiency of the estimates of the regression coefficients. Otherwise stated, the estimates of the standard errors of the regression coefficients will be larger, making it more difficult for any true effects in the equation to attain statistical significance. Hence, terms that have a value of zero in the population should be removed from the regression equation to permit more powerful tests of the other effects.

Researchers rarely know whether an interaction is zero in the population. Statistically, all the researcher can typically do to evaluate this hypothesis is to test the interaction in the sample and show it does not differ from zero. Unfortunately, as we will see in Chapter 8, tests of interactions often have low statistical power and may fail to detect small true interaction effects that exist in the population. This problem has led to conflicting recommendations in the literature. For example, in the ANOVA context, Cramer and Appelbaum (1980) have emphasized the increased efficiency that results when nonsignificant higher order terms are dropped from the model. They have argued that this gain more than compensates for the small amount of bias that may be introduced in the estimates of the lower order effects.[3] In contrast, Overall et al. (1981) focused on the problem of bias and showed under the conditions studied in their simulation that the use of the full model (equation 6.1) resulted in less bias, greater precision, and equal power relative to the use of the reduced model (equation 6.2) for tests of the lower order effects.

However, these results, while informative, should not necessarily be treated as being applicable in all contexts. As pointed out by Finney et al. (1984), "in actual research situations, as opposed to simulation analyses, the degree of bias for each approach depends on unknown—and, in many cases, unknowable—factors, such as the main and interactive effects of relevant variables that were not included in the model" (p. 91). Instead, Finney et al. (1984) argued that researchers should focus on the distinction we introduced above: the theoretical versus exploratory basis of the interaction. When there are strong theoretical expectations of an interaction, they conclude that the interaction should be retained in the final regression equation. Doing so informs the literature and leads to the accumulation of knowledge about the theory in question. Even when nonsignificant, the estimates of the effect size for the interaction may be combined across multiple studies through meta-analysis. Further, it is logically inconsistent to report the estimates of *constant* effects for X and Z from equation 6.2 when strong theory postulates an interaction.

In practice, estimates of the lower order effects derived from equation 6.1 versus from equation 6.2 will often be quite similar when all predictors have been centered. Indeed, Finney et al. (1984) note three cases in which they are identical. In each of these cases, there is no overlapping variance between the two predictors and their interaction. (a) If X and Z are uncorrelated with one another and are centered, then each is uncorrelated with the XZ term. (b) If X and Z are bivariate normal, then again $r_{X,XZ} = r_{Z,XZ} = 0$. (c) If the XZ pairs are balanced (i.e., for every $[-X, -Z]$ data point there is a corresponding $[X, Z]$ data point, and for every $[-X, Z]$ data point there is a corresponding $[X, -Z]$ data point), then again $r_{X,XZ} = r_{Z,XZ} = 0$. Although these exact conditions do not often hold in observed data, the $r_{X,XZ}$ and $r_{Z,XZ}$ correlations are often low with centered X and Z, so that estimates from equations 6.1 and 6.2 will be quite similar.

Recommendation

Our own recommendation concurs closely with that of Finney et al. (1984). In cases in which there are are strong theoretical grounds for expecting an interaction, the interaction, even if nonsignificant, should be retained in the final regression equation. Post hoc probing procedures can also be used as a supplemental guide to understanding how the interaction could potentially modify the overall results. Such probing is particularly important in cases in which the first order effects, though significant, are not large in magnitude and the statistical power of the test of the interaction is low. Conditional effects for the lower order terms from equation 6.1 may be reported where useful and appropriate, so long as centered X and Z have been used in the analysis and the nature of the effects is explained. However, in cases in which there is not a strong theoretical expectation of an interaction, step-down procedures should be used. The interaction should be dropped from the equation and the first order effects should be estimated using equation 6.2.

The remainder of this chapter focuses on a variety of global tests that can be useful in testing focused hyptheses about a variety of alternatives to the full regression model. We also present sequential step-down procedures that are useful in exploring complex regression equations. The analyst should always keep in mind the theoretical versus exploratory basis of the tests that are performed. This consideration informs the choice of data analytic strategy and the interpretation of the results. With Finney et al. (1984) we also strongly encourage researchers using these global and step-down testing procedures to report the findings of their preliminary analyses in order to place the main results in context.

Exploring Regression Equations Containing Higher Order Terms with Global Tests

There are many instances in which researchers may wish to use MR to conduct global tests of hypotheses. Such tests may be based on competing theoretical perspectives. For example, theory 1 may propose that only X and X^2 predict Y, whereas theory 2 may propose that Z and its interactions with X and X^2 also contribute to the prediction of Y. Or, the tests of the same hypotheses may be purely exploratory. Hypotheses about interactions and higher order effects may also be based on hunches about the data, prior data snooping, desires to identify or rule out possible moderators of an obtained effect, as well as many other possible sources. Such exploration is common in the case of orthogonal ANOVA. Researchers often explore the main and interactive effects of demographic factors (e.g., gender) in preliminary analyses in the absence of strong theoretical predictions. Complex factorial ANOVA models often include effects for which the researcher does not have specific hypotheses. For example, a complete, four factor ANOVA design generates a total of 15 effects—four main effects, six two-way interactions, four three-way interactions, and one four-way interaction—even though at most only a few of these effects have been predicted. Such exploratory analyses can yield new and interesting hypotheses that can then be tested in subsequent research. However, strong caution must be exercised in interpreting the results of the exploratory analyses within the original sample given the high experimentwise error rate (inflated alpha level) that results from conducting large numbers of tests. This problem can be minimized through the Bonferoni or other procedures that adjust the level of significance for the number of tests that have been conducted.

In the regression context, a variety of hypotheses about nonlinear and interactive effects may be tested through a general model comparison procedure. Model 1, the full model, contains all of the lower order terms plus the specific additional terms that are in question. Model 2, the reduced model, contains only the lower order terms. To compare the two models, a test of significance of the gain in prediction due to the inclusion of the additional terms is given as follows, with m, $n - k - 1$ df:

$$F = \frac{(R^2_{\text{in}} - R^2_{\text{out}})/m}{(1 - R^2_{\text{in}})/(n - k - 1)} \tag{6.5}$$

In this equation, R^2_{in} is the squared multiple correlation of the model containing the terms in question; R^2_{out} is the squared multiple correlation from the reduced model with the terms in question removed; m is the number of terms in the set of terms being explored; n is the number of cases; and k is the number of predictors in the full regression model, from which R^2_{in} is derived. We will use a number of variants of this general procedure throughout the remainder of the chapter.

As an illustration, if we wish to determine whether the Z and XZ terms contribute to the prediction in equation 6.1, we can compare the full model (equation 6.1) with a reduced model that eliminates X and XZ (equation 6.3). In this comparison, R^2_{in} is the multiple correlation from equation 6.1; R^2_{out} is that from equation 6.3; $m = 2$ for the b_2Z and b_3XZ terms in question; and $k = 3$, corresponding to the three predictors in equation

6.1. Of importance, all tests of the gain in prediction accounted for by a set of terms in a regression equation are *scale invariant*. This statement is true even when not all of the terms making up the predictor set are scale invariant. If this omnibus test is not significant, then all terms involved in the set are deleted (i.e., pooled into the error term) and the revised model is tentatively accepted.

Differences in the interpretation of the results and in the subsequent statistical procedure may appear at this point depending on the source of the reduced model. If the reduced model was based on substantive theory, it will be preferred over the alternative, full model on the grounds of parsimony. No further testing to eliminate terms from the reduced model would normally be conducted. In contrast, if the reduced model were exploratory, this revised model could now receive further scrutiny. In cases in which multiple terms remain in the reduced model, consideration would be given to the possibility of dropping additional terms from the reduced model. The purpose of such exploratory tests of complex regression models is to eliminate predictors for which there is no support and to simplify the model as much as is possible.

Some Global Tests of Models with Higher Order Terms

A variety of global tests may be made based on a single regression model. We will use the now familiar equation

$$\hat{Y} = b_1X + b_2X^2 + b_3Z + b_4XZ + b_5X^2Z + b_0 \qquad (6.6)$$

to illustrate several of the interpretable types of global tests that may be performed. The test chosen will ideally depend on substantive theory; in more exploratory cases, the test will depend on the specific questions that the researchers desire to answer.

1. Global Test of Linearity of Regression

Equation 6.6 may be contrasted to equation 6.2, $\hat{Y} = b_1X + b_2Z + b_0$, which contains only linear terms, to determine whether the regression model can be treated as being purely linear. If there is no significant gain in prediction from equation 6.2 to 6.6, then it is concluded that the simpler linear model of equation 6.2 is appropriate. Then the tests of b_1 and b_2 coefficients within equation 6.2 provide information about the significance of the linear effects of X and Z, respectively (see Darlington, 1990, p. 336; Pedhazur, 1982, p. 426).

2. Global Test of Curvilinearity of Regression

To determine whether the relationship of X to the criterion is curvilinear, the gain in prediction from equation 6.1 to equation 6.6 is tested. To see this, consider the reexpression of equation 6.6 in the form of a polynomial regression equation, given in equation 5.16 and reproduced here:

$$\hat{Y} = (b_1 + b_4 Z)X + (b_2 + b_5 Z)X^2 + (b_3 Z + b_0) \qquad (6.7)$$

The test of the gain in prediction from equation 6.1 to equation 6.6 provides a test of whether the linear combination of regression coefficients $(b_2 + b_5 Z)$ for the X^2 term is different from zero. In equations involving Z^2 terms, a similar reexpression of the regression equation in terms of Z and Z^2 provides the basis for a similar test evaluating whether the relationship of Z to the criterion is curvilinear.

3. Global Test of the Effect of One Variable

In equation 6.6, one may test whether there is any effect of one of the predictor variables, in either first order or higher order form. For a global test of variable X, equation 6.6 would be contrasted to equation 6.4, $\hat{Y} = b_2 Z + b_0$, which drops all terms containing X. If this test of gain in prediction is nonsignificant, it is concluded that variable X has no effect in the regression model. For a global test of variable Z, equation 6.6 would be contrasted to equation 6.8, which drops all terms containing Z:

$$\hat{Y} = b_1 X + b_2 X^2 + b_0 \qquad (6.8)$$

There are k such tests in a regression model based on k original predictors (here $k = 2$ corresponding to X and Z; see also Darlington, 1990, p. 336). Note that in these tests all interactions involving the variable in question are dropped, whereas interactions not involving the variable in question are retained.

4. Global Test of an ANOVA-Like Effect

As pointed out in Chapter 5, factors having more than 2 levels produce main effects and interactions in ANOVA with multiple degrees of freedom (df). For example, an ANOVA with 3 levels of X and 2 levels of Z yields an XZ interaction with 2 df. If these data were analyzed using MR, the interaction variance from the ANOVA would be accounted for by the XZ plus X^2Z terms in equation 6.6. Given this parallel structure, the same partitions of variation may be used in MR as in ANOVA. Thus a test of the overall XZ interaction analogous to that in ANOVA would result from comparing the full model represented by equation 6.6 with a reduced model in which the two terms involving XZ are dropped. This reduced model is represented by equation 6.9:

$$\hat{Y} = b_1X + b_2X^2 + b_3Z + b_0 \tag{6.9}$$

If there were no significant interaction, then both the b_4XZ and b_5X^2Z terms would be eliminated, leaving equation 6.9 as the appropriate equation.

Global tests of effects analogous to those in ANOVA may be employed in series. Suppose that the test of the interaction was not significant, yielding reduced regression equation 6.9. Equation 6.9 represents the main effect of X, represented by the X and X^2 terms, plus the Z main effect. Equation 6.9 would now be contrasted to equation 6.4 to determine the overall contribution of X to prediction. This is analogous to the test of a main effect of X in ANOVA where X has three levels. Finally, the linear effect of Z would be tested by contrasting equation 6.9 with equation 6.8.

The above tests of the overall X effect (X plus X^2) in the presence of Z and the Z effect in the presence of X and X^2 will be familiar to many ANOVA users. These procedures are identical to Appelbaum and Cramer's (1974) tests of main effects over and above other main effects. Appelbaum and Cramer (1974) referred to these tests with regard to the A and B main effects of an ANOVA as "A eliminating B" and "B eliminating A."

Appelbaum and Cramer (1974) also observed that in some instances the multiple correlation for an equation containing two main effects (e.g., equation 6.9) will be significant, but that neither individual effect will attain significance. They recommend further testing of the the X and Z main effects in which each effect is considered in a single equation. In the present case, the joint test of b_1 and b_2 in equation 6.8 would provide the test of X, and the test of b_2 in equation 6.4 would provide the test of Z. Such tests, termed "A ignoring B" and "B ignoring A" by Appelbaum and Cramer (1974), are recommended to clarify the impact of a single factor on a criterion. If only one of the pair of tests is significant, the associated predictor is taken as having an effect on (or association with) the criterion.

5. Global Test of the Equivalence of Two Regression Equations

Chapter 7 addresses the analysis of regression equations involving combinations of continuous and categorical predictor variables. In equation

6.1, let us assume that X is a continuous variable and Z is a dichotomous predictor variable representing two different groups (e.g., men versus women). The XZ interaction term in equation 6.1 then represents any difference in the slopes of the regression lines of Y on X between the two groups, here men versus women. Predictor Z represents differences in the Y scores between the two groups, evaluated at the mean of the full sample on the continuous variable.[4] If equation 6.1 is contrasted to equation 6.3, this provides a global test of whether there is any difference between the simple regression lines in the two groups (see Cohen & Cohen, 1983, pp. 312–313; Lautenschlager & Mendoza, 1986; Neter, Wasserman, & Kutner, 1989, p. 368).

Structuring Regression Equations with Higher Order Terms

In all of these global tests, the equations containing higher order terms are deliberately structured to contain all lower order terms of which the higher order terms are comprised. Tests of the contributions of higher order terms should consider prediction of these terms over and above the lower order terms. This is because higher order terms actually represent the effects they are intended to represent *if and only if* all lower order terms are partialed from them (Cohen, 1978). The XZW term covered in detail in Chapter 4 only represents the linear × linear × linear component of the XZW interaction if all lower order effects (i.e., X, Z, W, XZ, XW, and ZW) have been partialed from the XZW term. A global test of the curvilinearity of X in equation 6.6 would not be accomplished by determining the significance of the multiple correlation in the equation $\hat{Y} = b_2X^2 + b_5X^2Z + b_0$. In such an equation the higher order curvilinear effects would be confounded with portions of the linear X, linear Z, and the linear by linear XZ interaction. In sum, we recommend strongly that in structuring regression equations with higher order terms all lower order terms be included (see Allison, 1977; Cleary & Kessler, 1982; Cohen & Cohen, 1983; Darlington, 1990; Pedhazur, 1982; Peixoto, 1987; Stone & Hollenbeck, 1984 for further discussion of the necessity for the inclusion of lower order terms when higher order terms are tested).

Readers will occasionally encounter models in research literature in which lower order terms have been omitted. Fisher (1988) described a theoretical model in which one predictor X has a direct effect on the criterion Y; a second predictor Z modifies the effect of X on Y, but has no direct effect on Y. This theorizing led to the regression equation $\hat{Y} = b_1X$

$+ b_3XZ + b_0$, with the b_2Z term for the direct effect of X on Y omitted. We suggest that it is usual with such theorizing to test the direct effect of X on Z and to show lack of support for its operation, rather than to omit it from the model. Hence, we advocate the use of models with all lower order effects included for theory testing. Demonstrations that effects predicted to be nonexistent by theory do not accrue are valuable in theory building.

Sequential Model Revision of Regression Equations Containing Higher Order Terms: Exploratory Tests

There are two instances in which the examination of complex regression equations will proceed one term at a time. First, in tests of exploratory hypotheses, if a global test involving several terms is significant, then the individual terms in the set are tested to characterize further the nature of the effect studied in the global test. The second instance is in the absence of global tests. One may adopt a strategy of sequentially exploring a complex regression equation term by term without having first performed more global tests.

We showed in Chapter 3 that all regression coefficients except that for the highest order term(s) are scale dependent, and that their significance levels will vary widely with changes in scaling of the first order predictors. Thus what is required is an approach to model exploration using sequential term-by-term testing that involves tests only of scale-invariant terms at each stage of analyses. A step-down hierarchical examination of the regression equation with model respecification after each step satisfies this requirement. The approach begins with the full equation; nonsignificant terms are then omitted sequentially in stages beginning with the highest order term in the equation. At each step the scale-invariant terms are identified and only these terms are tested for statistical significance. Such an approach is outlined for equation 6.6 by Peixoto (1987).

In any regression equation a term will be scale free if the complete combination of letters and superscripts of the term is not included in any other term. In equation 6.1 the XZ term is scale free; in equation 6.6 it is not scale free because it is included in the X^2Z term. Likewise, X^2 is scale free in equation 6.9 but not in equation 6.6. Appendix B presents an algorithm for identifying scale-free terms in regressions through the complexity of equations 4.1 and 5.5, containing XZW or X^2Z^2. It also shows algebraically that the terms so identified are scale free. A simple, alternative computer-based procedure that is useful for complex regression equations is to estimate the regression equation twice. In the first run, the original values of the predictor variables are used; in the second run, a constant is added to each of the predictor variables (e.g., $X^* = X + 10$). Those terms in the regression equation that do not change between the two runs are scale invariant. Note that each time a term is dropped from the equation in model revision, new terms become scale free, so that this computer checking method would have to be repeated at each step.[5]

Application of Sequential Testing Following a Global Test

If a global test involving more than one term is significant, then the terms involved in the global test should be explored individually beginning with the highest order term of the set, the only term that is scale invariant. If nonsignificant, the highest order term is eliminated and the model is revised. Those remaining terms of the global test set that are scale free in the revised equation are then tested for significance. For example, we described on page 107 a global test of the linearity of regression that contrasted equations 6.6 and 6.2. If this global test were significant, then the terms of the set X^2, XZ, and X^2Z would be candidates for follow-up tests. The X^2Z term would be tested for significance first, by testing the significance of the b_5 coefficient in equation 6.6, or equivalently by testing the gain in prediction of equation 6.6 over equation 6.10:

$$\hat{Y} = b_1X + b_2X^2 + b_3Z + b_4XZ + b_0 \qquad (6.10)$$

If the b_5 coefficient were significant, this interaction would be probed following the prescriptions of Chapter 5, Case 4a. If the coefficient were nonsignificant and were dropped from the equation, then in resulting equation 6.10, there are two scale-independent terms, b_2X^2 and b_4XZ. Each of these terms is tested separately. The b_4XZ term is tested by contrasting equation 6.10 to equation 6.9; the b_2X^2 term is tested by contrasting equation 6.10 to equation 6.11, presented below.

$$\hat{Y} = b_1X + b_3Z + b_4XZ + b_0 \qquad (6.11)$$

Note that both equations 6.9 and 6.10 are hierarchically "well formulated" (Peixoto, 1987) in that all lower order terms are represented. Suppose further that the b_2X^2 term were significant, whereas the b_4XZ term were not. Then equation 6.9 would be retained. The source of the deviation from linearity found in the global test would have been identified as resulting from a curvilinear regression of Y on X. This relationship would then be further characterized using the strategies for probing curvilinear relationships described in Chapter 5, Case 2.

General Application of Sequential Testing

An alternative strategy for exploring complex regression equations is to simplify the model on a term-by-term basis without first having performed global tests. The researcher simply begins with the highest order term in the regression equation and steps down through the hierarchy following the algorithm for scale-independent terms. The procedure stops when all nonsignificant higher order terms are eliminated from the equation. If the final reduced equation still contains higher order terms, then these terms should be probed using the prescriptions from Chapters 2 through 5, or from Chapter 7 if combinations of discrete and continuous predictors are involved.

Present Approach Versus that Recommended by Cohen (1978)

Cohen (1978) recommended a hierarchical step-up approach to examining regression equations that include higher order terms. In his approach, lower order effects are tested in equations containing only these effects; interactions are then tested for their contribution over and above main effects. Cohen's approach handles the scale invariance problem as does the step-down approach in that lower order scale-dependent terms are never tested in the presence of higher order terms. However, Cohen's step-up approach can lead to the interpretational problem of considering "main effects" before the analyst has determined whether interactions exist. Indeed, investigators have misused the step-up approach by interpreting the coefficients from an equation containing only first order effects as main effects when the subsequent test showed the interaction to be significant. The step-down approach handles the scale invariance problem and also helps insure that lower order effects will be interpreted as conditional or average effects once a higher order effect has been shown to exist. The step-down hierarchical approach is consistent with the position taken by Cohen that first order terms should not be tested if there is a significant interaction.

Variable Selection Algorithms

The familiar automatic stepwise forward (build-up) and stepwise backward (tear-down) variable selection algorithms should not be confused with the hierarchical step-up and step-down procedures discussed here. The selection of predictors in stepwise procedures available in standard statistical packages is based solely on the predictive utility of individual predictors over and above other predictors. Their use in the context of complex regression equations containing higher order terms will lead to reduced regression equations that are not hierarchically well-formulated, that is, in which all necessary lower order terms are included. The identical problem holds for the all-possible-subset regression algorithms in which all possible regression equations containing 1 through k predictors from a set of k predictors are generated and the ''best'' equation in terms of predictive utility is selected. With regression equations containing higher order terms, none of typical automatic search procedures is appropriate. Only a procedure that preserves the hierarchy of variables at each stage should be employed (see also Peixoto, 1987).

Summary

This chapter intially considers two questions stemming from the nonindependence of terms in regression equations containing interactions. First, the interpretation of the lower order coefficients in the presence of an interaction is reviewed. Second, the trade-off between bias and efficiency in the tests of regression coefficients involved in dropping nonsignificant terms from a regression equation is discussed. The important role of the theoretical versus exploratory basis of the interaction is emphasized in the choice of testing procedure and in the interpretation of the results. A variety of global step-down tests of focused hypotheses (e.g., the presence of curvilinearity, the overall effect of a single variable) are introduced. A term-by-term strategy is presented for exploring complex regression equations through the step-down elimination of nonsignificant higher order predictors. Methods for the identification of scale-invariant terms in any regression equation with higher order terms are provided. An advantage of a hierarchical step-down procedure over step-up procedures is presented.

Notes

1. The percentage of variance uniquely shared between the predictor and the criterion in each regression equation is the square of the standardized regression coefficent.

2. An infrequently used alternative procedure is for the analyst to specify the order of tests in a hierarchical step-up procedure. The order of tests is based on strong theory, the temporal precedence of the predictor variables, or both. For example, in a study of the effect of students' race (X), high school GPA (Z), and their interaction (XZ) on college GPA, the researcher could argue that race precedes high school GPA and thus accounts for some of the variance in high school GPA. Race and high school GPA could also (though much more weakly) be argued to precede the race $\times$ GPA interaction. Under this strong set of assumptions, the test of b_1 in equation 6.3 would provide the test of the race effect, the test of b_2 in equation 6.2 would provide the test of effect of high school GPA, and the test of b_3 in equation 6.1 would provide the test of the race $\times$ high school GPA interaction. In this strategy, all of the variance shared between race, high school GPA, and their interaction is apportioned to race; the variance shared only between high school GPA and the interaction is apportioned to high school GPA. The test of b_3 is once again a test of the unique variance of the interaction. In the absence of a strong theoretical claim that X causes Z, the attribution of the shared variance to X cannot be logically justified. Thus claims for the validity of this approach are strongly dependent on the judged adequacy of the underlying substantive theory.

3. Readers wishing more advanced treatments of the bias versus efficiency issue in sequential testing (pretest estimators) should consult Judge, Hill, Griffiths, Lutkepul, and Lee (1982, particularly Chapter 21) and Judge and Bock (1978).

4. As will be explained in Chapter 7, this interpretation assumes that the group variable has been dummy coded. If effect coding has been used, this effect represents the difference of each group from the unweighted mean, again evaluated at the mean of the continuous variable.

5. We thank David Kenny for suggesting the computer test for scale invariance.

7 Interactions Between Categorical and Continuous Variables

Thus far we have focused on the treatment of interactions between continuous predictor variables. We now consider problems in which categorical predictor variables having two or more levels interact with continuous predictor variables. We also discuss techniques for post hoc probing to aid in the interpretation of significant interactions involving categorical and continuous variables.

Coding Categorical Variables

A number of methods for coding categorical variables have been proposed (see e.g., Cohen & Cohen, 1983; Darlington, 1990; Pedhazur, 1982). Two methods are considered here: (a) dummy variable coding and (b) unweighted effects coding.

Dummy Variable Coding

Dummy variable coding is the most frequently utilized procedure in the literature for representing categorical variables in regression equations. Contrary to our recommendations for continuous variables, the dummy coding procedure does not center the comparisons involving the categorical variable(s). Nonetheless, as will be seen below, the results of this procedure are easily interpretable.

To illustrate the use of dummy coding, imagine a researcher is studying the starting salaries of bachelor degree graduates in three colleges: Liberal Arts (LA), Engineering (E), and Business (BUS). College is the categorical variable in this example. One of the colleges (e.g., LA) is designated as the comparison group; this designation may be arbitrary, based on theory, or because of a special interest in comparing the other two colleges with this baseline. In general, $G - 1$ dummy variables will be needed, where G is the number of groups (levels of the categorical variable). With three levels of college, $3 - 1 = 2$ dummy variables will be needed. Three possible sets of dummy variable codes for comparing these three colleges are presented in Table 7.1.

Table 7.1
Three Dummy Variable Coding Systems for College Data

a. LA as Comparison Group			b. E as Comparison Group			c. BUS as Comparison Group		
	D_1	D_2		D_1	D_2		D_1	D_2
LA	0	0	LA	1	0	LA	1	0
E	1	0	E	0	0	E	0	1
BUS	0	1	BUS	0	1	BUS	0	0

Throughout this section we will use the first set of dummy variable codes depicted in Table 7.1a. In this coding system, the first dummy variable (D_1) compares E with the LA comparison group which is assigned a value of 0. The second dummy variable (D_2) compares BUS with the LA comparison group. In dummy coding, (a) the comparison group is assigned a value of 0 in all dummy variables, (b) the group being contrasted to the comparison group is assigned a value of 1 for that dummy variable only, and (c) groups not involved in the contrast are also assigned a value of 0 for that dummy variable. Note that dummy codes are partial effects that are conditioned on all $G - 1$ dummy variables being present in the regression equation.[1]

Continuing with our example, suppose the researcher has sampled a small number ($N = 50$) of university graduates and has recorded their college, grade point average, and starting salary. College and GPA are the predictor variables and starting salary is the outcome variable of interest. These data are presented in Table 7.2. As in Chapter 5, we will consider the interpretation of a series of regression equations of increasing complexity.

Table 7.2
Hypothetical Data: Starting Salaries in Three Colleges

Sub. No.	College	GPA	Salary ($)	Sub. No.	College	GPA	Salary ($)
1	LA	2.54	21,140	26	E	2.18	28,219
2	LA	2.25	20,667	27	E	1.93	27,946
3	LA	2.69	21,003	28	E	2.31	28,053
4	LA	2.84	21,269	29	E	2.45	28,209
5	LA	2.73	20,831	30	E	2.35	27,899
6	LA	2.83	21,370	31	E	2.44	28,295
7	LA	2.48	20,435	32	E	2.13	27,672
8	LA	2.58	20,584	33	E	2.22	27,756
9	LA	3.95	21,604	34	E	3.41	28,065
10	LA	3.00	20,937	35	E	2.58	27,885
11	LA	2.59	20,625	36	BUS	2.78	23,942
12	LA	2.27	20,389	37	BUS	2.50	23,205
13	LA	3.14	21,490	38	BUS	2.92	23,962
14	LA	2.95	21,007	39	BUS	3.08	24,369
15	LA	2.67	21,063	40	BUS	2.96	23,840
16	LA	2.67	21,003	41	BUS	3.06	24,452
17	LA	2.67	20,586	42	BUS	2.72	23,218
18	LA	2.89	21,084	43	BUS	2.82	23,455
19	LA	2.94	21,256	44	BUS	4.00	25,790
20	LA	3.43	21,651	45	BUS	3.22	24,206
21	LA	2.75	20,794	46	BUS	2.83	23,506
22	LA	1.93	20,380	47	BUS	2.52	22,961
23	LA	3.04	20,961	48	BUS	3.36	24,868
24	LA	3.13	21,796	49	BUS	3.18	24,223
25	LA	3.05	21,075	50	BUS	2.91	24,004

	GPA Mean	Salary Mean ($)
LA	2.80	21,000
E	2.40	28,000
BUS	3.00	24,000

NOTE: LA = Liberal Arts; E = Engineering; BUS = Business Administration.
GPA is on a 4-point system where A = 4.0, B = 3.0, C = 2.0, D = 1.0, F = 0.0.

Categorical Variable Only

Consider the simple regression equation containing only the dummy variables:

$$\hat{Y} = b_1 D_1 + b_2 D_2 + b_0 \tag{7.1}$$

This equation compares the mean starting salaries (Y) of graduates of the three colleges. Let us substitute the dummy codes for each college from Table 7.1a into this equation.

$$\text{LA:} \quad \hat{Y} = b_1(0) + b_2(0) + b_0 = b_0$$

$$\text{E:} \quad \hat{Y} = b_1(1) + b_2(0) + b_0 = b_1 + b_0$$

$$\text{BUS:} \quad \hat{Y} = b_1(0) + b_2(1) + b_0 = b_2 + b_0$$

From these substitutions, we see that b_0 represents the mean (predicted) value of Y for the comparison group, here the LA graduates. $b_0 + b_1$ represents the mean value of Y for the E graduates, and $b_0 + b_2$ represents the mean value of Y for the BUS graduates. Table 7.3a(i) presents the results of the analysis for the college salary data. The estimates for the three b coefficients are \$21,000 for b_0, \$6,999.90 for b_1, and \$3,000.10 for b_2. As can be seen in Table 7.2, b_0 = mean for LA graduates, $b_1 + b_0$ = mean for E graduates (\$27,999.90), and $b_2 + b_0$ = mean for BUS graduates (\$24,000.10). These data are depicted in Figure 7.1a.

The joint test of b_1 and b_2 (see Chapter 6) compares the mean starting salaries of the three colleges and is equivalent to a one-way ANOVA. The $SS_{\text{regression}}$ (predictable sum of squares) from the regression analysis in Table 7.3a(i) is exactly equal to $SS_{\text{treatment}}$ in the one-way ANOVA. The F-tests from the two analyses are equivalent as well. The results of the joint test in the regression analysis show significant differences among the groups. The test of b_1 compares the means for the E and LA groups; the test of b_2 compares the means for the BUS and LA groups. Both contrasts are significant.[2] (See also note 1, p. 138).

Categorical and Continuous Variables

We now add an effect of GPA on starting salary to the equation, resulting in

$$\hat{Y} = b_1 D_1 + b_2 D_2 + b_3 \,\text{GPA} + b_0 \tag{7.2}$$

As usual in our treatment of continuous variables, we will center GPA throughout this section:

$$\text{GPA} = \text{Original GPA} - \text{Mean GPA for entire sample}$$

Table 7.3
Analyses of Progression of Regression Equations

a. Dummy Variable Coding

(i) Test of dummy variables only
$\hat{Y} = b_1 D_1 + b_2 D_2 + b_0$
$\hat{Y} = (6{,}999.9)(D_1) + (3{,}000)(D_2) + 21{,}000$
Joint test of b_1, b_2: $R^2 = 0.969$, $F(2, 47) = 744$, $p < .001$
Test of b_1: $t(47) = 38.0$, $p < .001$
Test of b_2: $t(47) = 18.7$, $p < .001$
$SS_{reg} = 360{,}492{,}000$; $SS_{res} = 11{,}386{,}717$

(ii) Test of dummy variables and continuous variable
$\hat{Y} = b_1 D_1 + b_2 D_2 + b_3 \text{GPA} + b_0$
$\hat{Y} = (7{,}377)(D_1) + (2{,}821)(D_2) + 943(\text{GPA}) + 20{,}978$
Joint test of b_1, b_2, b_3: $R^2 = 0.987$, $F(3, 46) = 1{,}124$, $p < .001$.
Test of b_3: $R^2_{change} = 0.017$, $F(1, 46) = 58.62$, $p < .001$.
$SS_{reg} = 366{,}872{,}239$; $SS_{res} = 5{,}006{,}479$

(iii) Test of dummy variables, continuous variable, and interaction
$\hat{Y} = b_1 D_1 + b_2 D_2 + b_3 \text{GPA} + b_4(D_1 \times \text{GPA}) + b_5(D_2 \times \text{GPA}) + b_0$
$\hat{Y} = 7{,}065(D_1) + (2{,}619)D_2 + (790)\text{GPA} + (-667)(D_1 \times \text{GPA})$
$+ (1{,}082)(D_2 \times \text{GPA}) + 20{,}982$
Joint test of b_1–b_5: $R^2 = 0.994$, $F(5, 44) = 1{,}412$, $p < .001$.
Joint test of b_4, b_5: $R^2_{change} = 0.007$, $F(2, 44) = 25.80$, $p < .001$.
Test of b_3: $t(44) = 6.76$, $p < .001$
Test of b_4: $t(44) = -2.98$, $p < .01$
Test of b_5: $t(44) = 5.33$, $p < .001$
$SS_{reg} = 369{,}574{,}652$; $SS_{res} = 2{,}304{,}066$

Once again we substitute the dummy codes for each college into the equation:

$$\text{LA:} \quad \hat{Y} = b_3 \text{ GPA} + b_0$$

$$\text{E:} \quad \hat{Y} = b_1 + b_3 \text{ GPA} + b_0$$

$$\text{BUS:} \quad \hat{Y} = b_2 + b_3 \text{ GPA} + b_0$$

As can be seen, equation 7.2 implies that each college is represented by a separate regression line, with each line having an identical slope, b_3. This means the three regression lines will be parallel to one another, as is illustrated in Figure 7.1b. b_0 represents the predicted salary value for

Table 7.3, continued

b. Unweighted Effects Coding

(i) Test of effects variables only
$\hat{Y} = b_1 E_1 + b_2 E_2 + b_0$
$\hat{Y} = \$3,666.58(E_1) + (-333.26)(E_2) + \$24,333.32$
Joint test of b_1, b_2: $R^2 = 0.969$, $F(2, 47) = 744$, $p < .001$
Test of b_1: $t(47) = 31.4$, $p < .001$
Test of b_2: $t(47) = -3.2$, $p < .01$
$SS_{reg} = 360{,}492{,}000$; $SS_{res} = 11{,}386{,}718$

(ii) Test of effect variables and continuous variable
$\hat{Y} = b_1 E_1 + b_2 E_2 = b_3 GPA + b_0$
$\hat{Y} = 3,978.1(E_1) + (-578.7)(E_2) + 943(GPA) + 24,377.7$
Joint test of b_1, b_2, b_3: $R^2 = 0.987$, $F(3, 46) = 1,124$, $p < .001$
Test of b_3: $R^2_{change} = 0.017$, $F(1, 46) = 58.62$, $p < .001$
$SS_{reg} = 366{,}872{,}239$; $SS_{res} = 5{,}006{,}479$

(iii) Test of effect variables, continuous variables, and interaction
$\hat{Y} = b_1 E_1 + b_2 E_2 + b_3 GPA + b_4(E_1 \times GPA) + b_5(E_2 \times GPA) + b_0$
$\hat{Y} = (3,836.9)E_1 + (-609.2)E_2 + 928.3\ GPA$
$+ (-805.4)(E_1 \times GPA) + (943.7)(E_2 \times GPA) + 24,209.3$
Joint test of b_1–b_5: $R^2 = 0.994$, $F(5, 44) = 1,412$, $p < .001$
Joint test of b_4, b_5: $R^2_{change} = 0.007$, $F(2, 44) = 25.80$, $p < .001$
Test of b_3: $t(44) = 10.00$, $p < .001$
Test of b_4: $t(44) = -5.60$, $p < .01$
Test of b_5: $t(44) = 7.08$, $p < .001$
$SS_{reg} = 369{,}574{,}652$; $SS_{res} = 2{,}304{,}066$

LA graduates having the GPA variable equal to 0. Because we have centered GPA, this corresponds to the mean GPA of the entire sample.[3] In Table 7.3a(ii) note that the value of b_0 (\$20,978) has changed from that in our first equation (\$21,000). b_0 now represents the predicted starting salary of LA graduates at the mean value of GPA for the entire sample (2.78) in contrast to its earlier interpretation as the mean GPA for the LA graduates.

b_1 represents the distance between the LA and E regression lines, and b_2 represents the distance between the BUS and LA regression lines. b_1 and b_2 represent the difference between mean starting salaries for E and LA graduates and BUS and LA graduates, respectively, but now adjusted for (conditioned on) GPA. The differences are now attributable to the differences between the colleges above and beyond (independent of) differences in GPA between the colleges. These changes in the meaning of

a. $\hat{Y} = b_1 D_1 + b_2 D_2 + b_0$

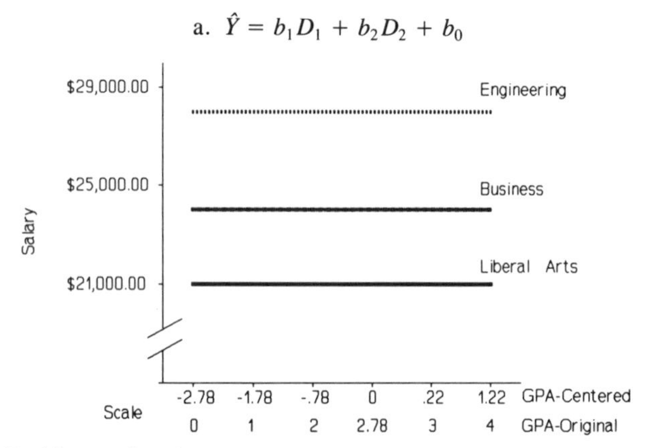

NOTE: In this equation the predicted salary is equal to the mean for the college regardless of the level of GPA.

b. $\hat{Y} = b_1 D_1 + b_2 D_2 + b_3 \text{ GPA} + b_0$

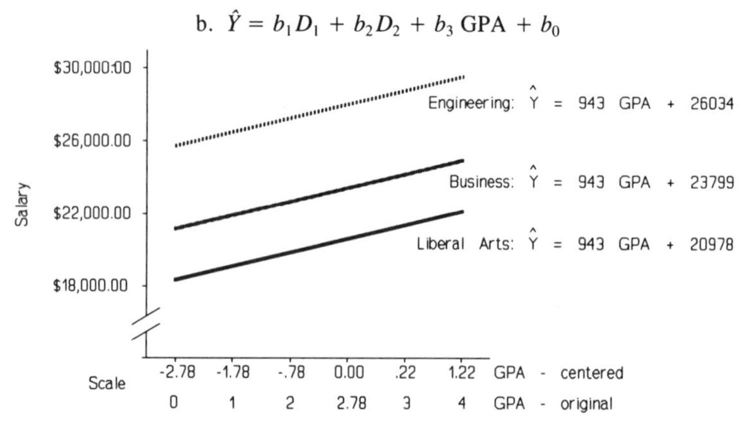

NOTE: Simple regression lines for the three colleges are parallel.

Figure 7.1. Simple Regression Lines for Three Colleges

the b_1 and b_2 coefficients are reflected in the changed values of these coefficients seen in Table 7.3a, (i) and (ii). b_1 now equals \$7,377 and b_2 now equals \$2,821.

The b_3 coefficient represents the slope relating GPA to starting salary. Recall in equation 7.2 we have assumed this relationship to be identical in each of the colleges. b_3 = \$943 indicates that graduates could expect a change in starting salary of \$943 if their GPA changed 1.0 grade point. For example, an increase in GPA from 2.78 to 3.78 would lead to a

c. $\hat{Y} = b_1 D_1 + b_2 D_2 + b_3 \text{ GPA} + b_4 D_1 \text{ GPA} + b_5 D_2 \text{ GPA} + b_0$

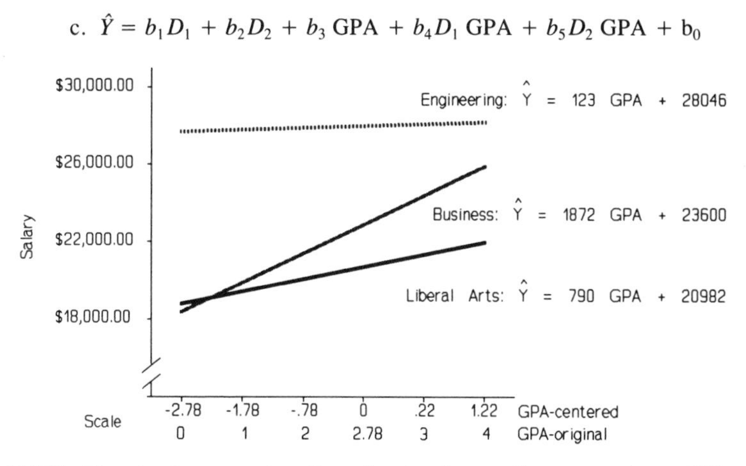

NOTE: The simple regression lines for the three colleges now have different slopes.

Figure 7.1, continued

predicted increase in starting salary of $943. Note that in equation 7.2 the effects of college are statistically removed from the regression coefficient b_3 for GPA. If this were not done (i.e., D_1 and D_2 were not included in the equation), b_3 would equal $-\$1{,}080$, reflecting the relatively low GPAs and high salaries of the engineers.

Categorical and Continuous Variables and their Interaction

Finally, let us consider a regression equation in which the slopes representing the relationship between GPA and starting salary are permitted to differ among the colleges. This means that the regression lines may not be parallel, indicating a potential interaction between the categorical and continuous variables. In general, the interaction between a continuous and a categorical variable is formed by multiplying the continuous variable by each of the dummy variables comprising the categorical variable. In the present example, the following equation adds two terms to represent the interaction:

$$\hat{Y} = b_1 D_1 + b_2 D_2 + b_3 \text{ GPA} + b_4 (D_1 \times \text{GPA}) + b_5 (D_2 \times \text{GPA}) + b_0 \tag{7.3}$$

To understand the meaning of these equations, we substitute the dummy codes for each of the colleges into this equation. This substitution produces three *simple regression equations*, analogous to those derived in Chapters 2 through 5. Terms in which the value of either dummy variable equals 0 are omitted.

$$\text{LA:} \quad \hat{Y} = b_3\,\text{GPA} + b_0$$

$$\text{E:} \quad \hat{Y} = b_1(1) + b_3\,\text{GPA} + b_4\,\text{GPA} + b_0$$
$$= (b_1 + b_0) + (b_3 + b_4)\,\text{GPA}$$

$$\text{BUS:} \quad \hat{Y} = b_2(1) + b_3\,\text{GPA} + b_5\,\text{GPA} + b_0$$
$$= (b_2 + b_0) + (b_3 + b_5)\,\text{GPA}$$

This substitution makes it clear that each college now has its own linear regression line, with each line having a separate intercept and slope. The three regression lines corresponding to each of the colleges are depicted in Figure 7.1c. b_3 represents the slope of the line for the LA graduates, $b_3 + b_4$ represents the slope of the line for the E graduates, and $b_3 + b_5$ represents the slope of the line for the BUS graduates. From Table 7.3a(iii), b_3 = \$790 is the slope for the LA graduates, $b_3 + b_4$ = \$123 is the slope for the E graduates, and $b_3 + b_5$ = \$1,872 is the slope for the BUS graduates.

b_0 represents the intercept of the LA regression line, evaluated in the present centered case at the mean of the entire sample of 50 students. b_1 represents the distance between the LA and E regression lines and b_2 represents the distance between the LA and BUS regression lines, both distances evaluated at the mean GPA of the entire sample. From Table 7.3a(iii), we see that these values are b_0 = \$20,982, b_1 = \$7,065, and b_2 = \$2,619. (Note that the intercept for the LA group is b_0 = \$20,982, for the E group it is $b_1 + b_0$ = 7,064.71 + 20,981.52 = \$28,046, and for the BUS group it is $b_2 + b_0$ = 2,618.57 + 20,981.52 = \$23,600.09.) These coefficients do not equal the values obtained in estimating equation 7.2, which does not contain the interaction terms: Equation 7.2 attributes a portion of the interaction variance to the lower order terms. Note that the distance estimates are only meaningful when interpreted at the mean GPA of the sample, because the regression lines are not parallel.

To provide another perspective on these slope and intercept estimates, we separately computed a simple linear regression of starting salary on GPA (centered) within each college. The results were as follows:

$$\text{LA:} \quad \hat{Y} = 790 \text{ GPA} + 20{,}982$$

$$\text{E:} \quad \hat{Y} = 123 \text{ GPA} + 28{,}046$$

$$\text{BUS:} \quad \hat{Y} = 1{,}872 \text{ GPA} + 23{,}600$$

Note that the estimates for the slope and intercept within each college are, within rounding error, equal to the values reported above based on a model containing the dummy variables, the continuous variable, and the interactions above.

A joint test of b_4 and b_5 provides the overall test of the statistical significance of the interaction. As can be seen in Table 7.3a(iii), this test is significant as are the individual tests of each of the contrasts involving a dummy variable in interaction with the continuous variable, GPA.

Finally, the point of intersection of any pair of the lines can be calculated to determine whether the lines cross within the useful range of the continuous variable, here GPA. Each pair of lines may cross at a different point. To calculate the point of intersection, the equations for the two regression lines are set equal to each other and solved for the continuous variable (GPA). For the regression lines representing the LA and E groups above,

$$b_3 \text{ GPA} + b_0 = b_1 + b_3 \text{ GPA} + b_4 \text{ GPA} + b_0$$

So the intersection point for the LA and E lines is at GPA $= -b_1/b_4 = -7{,}065/-667 = 10.59$ (see Table 7.3a(iii) for values of b_1 and b_4). Setting the equations equal for the other possible pairs of groups produces two additional intersection points: the intersection of LA and BUS is -2.42 and the intersection of BUS and E is 2.54. Recall that we have centered GPA. To place these values back in the original metric, the sample mean must be added to these values leading to the following intersection points on the original GPA scale: LA–E = 13.36, LA–BUS = 0.35, and E–BUS = 5.31. Figure 7.1c illustrates the three simple regression lines corresponding to each college.

As discussed previously in Chapter 2, interactions in which all of the points of intersection fall outside of the useful range of the continuous variable are termed *ordinal*, whereas interactions in which at least one of the points of intersection fall within the useful range of the continuous variable are termed *disordinal*. Two of the three points of intersection fall outside the possible range of GPA; however, the LA–BUS intersection (0.35) does fall within the theoretical 0.0–4.0 range of GPA. Note, however, that the LA–BUS intersection point does fall outside of the range of GPAs observed in our sample and very likely represents an impossible GPA for an actual graduate. Hence, the LA–BUS interaction should also be considered to be ordinal.

More generally, the point of intersection on the continuous variable can be calculated using the slopes and intercepts of the two regression lines according to the following equation:

$$\text{Intersection point} = \frac{I_1 - I_2}{S_2 - S_1} \qquad (7.4)$$

In this equation, I_1 is the intercept for group 1, I_2 is the intercept for group 2, S_1 is the slope for group 1, and S_2 is the slope for group 2.

Higher Order Effects and Interactions

As we saw in Chapter 5 with continuous variables, higher order terms can also be added to the equation. Because the recommendations of Chapter 5 can be applied here, we will consider only briefly two examples.

As a first example, the potential linear and quadratic effects of GPA on starting salary, each of which are assumed to be identical across groups, can be examined by adding a GPA^2 term to equation 7.2. This results in equation 7.5:

$$\hat{Y} = b_1 D_1 + b_2 D_2 + b_3\,\text{GPA} + b_4\text{GPA}^2 + b_0 \qquad (7.5)$$

Extending the example further, if the linear and quadratic effects of GPA are both permitted to differ among the groups, then additional terms must be added to equation 7.5, resulting in equation 7.6:

$$\begin{aligned}\hat{Y} = b_1 D_1 &+ b_2 D_2 + b_3\,\text{GPA} + b_4\text{GPA}^2 + b_5(D_1 \times \text{GPA}) \\ &+ b_6(D_2 \times \text{GPA}) + b_7(D_1 \times \text{GPA}^2) \\ &+ b_8(D_2 \times \text{GPA}^2) + b_0 \qquad (7.6)\end{aligned}$$

Equation 7.6 permits both the linear and quadratic components of the GPA–starting salary relationship to differ among the three colleges. Comparison of equation 7.6 with one in which the ($D_1 \times \text{GPA}^2$) and ($D_2 \times \text{GPA}^2$) terms are not included permits a test of the significance of the quadratic component of the college × GPA interaction.

The point(s) of intersection for the two groups can be determined by setting the equations for the two groups to be equal and solving for the continuous variable. For example, substitution into equation 7.6 and algebraic manipulation shows that the LA and E lines will cross at

$$\text{Intersection point 1} = \frac{-b_4 + [b_4^2 - 4(b_1)(b_7)]^{1/2}}{2(b_1)}$$

$$\text{Intersection point 2} = \frac{-b_4 - [b_4^2 - 4(b_1)(b_7)]^{1/2}}{2(b_1)}$$

If the solutions for intersection points 1 and 2 are equal, there is only one intersection point. None, one, or both of the intersection points may occur within the useful range of the continuous variable.

In summary, just as was the case for continuous variables, higher order terms can be added to the regression equation to test specific hypotheses. Each of the dummy variables representing the categorical variable must be included as first order terms and in all interactions. As with continuous variables, all of the lower order terms involved in interactions must be included in the equation. Quadratic (or other higher order) functions of dummy variables are never included in the regression equation, as they do not lead to interpretable effects.

Unweighted Effects Coding

Unweighted effects coding is a second, relatively frequently used coding system. In this coding system the two groups involved in the contrast specified by the code are compared with the *unweighted* mean of all of the groups. As was defined in Chapter 2, the comparison group is always coded -1, the group being contrasted to the comparison group is assigned a value of 1, and groups not involved in the contrast are assigned a value of 0. Returning to our college-salary example, we again arbitrarily designate LA as the comparison group. Two variables (effects codes) are again needed to contrast LA to E and BUS, respectively. Table 7.4 provides the unweighted effects codes for these contrasts. Substituting these

Table 7.4
Unweighted Effects Codes for College Example

	E1	E2
LA	−1	−1
E	1	0
BUS	0	1

values into equation 7.1, we obtain the following set of equations for each of the three colleges:

$$\text{LA:} \quad \hat{Y} = -b_1 - b_2 + b_0$$

$$\text{E:} \quad \hat{Y} = b_1 + b_0$$

$$\text{BUS:} \quad \hat{Y} = b_2 + b_0$$

In these equations, b_0 represents the unweighted grand mean of the three groups, that is, b_0 = [Mean(LA) + Mean(E) + Mean(BUS)]/3 = [$21,000 + $27,999.9 + $24,000.1]/3 = $24,333.3. b_1 represents the deviation of the mean of the group labeled 1 in the E_1 contrast, here Mean(E), from the unweighted grand mean. This value is $27,999.9 − $24,333.3 = $3,666.6. b_2 represents the deviation of the mean of the group labeled 1 in the E_2 contrast (BUS) from the unweighted grand mean.

Table 7.3b presents the results for the unweighted effects code analysis for several of the regression equations discussed above. A comparison of these results with those using dummy coding show both similarities and differences in the results. (a) Overall tests of R^2 or change in R^2 corresponding to adding a categorical, continuous, or interaction variable are identical across the two coding systems. (b) Tests of the coefficients of the continuous variable (b_3) and each term of the interaction (b_4, b_5) are different.[4] (c) Tests of the intercept and the dummy (or unweighted effect) variables differ. These differences reflect the dissimilar interpretations of the b coefficients in the two coding systems. In dummy variable coding, the contrasts are with the comparison group; in unweighted effects coding, the contrasts are with the unweighted mean of the sample.

To illuminate further the difference in interpretations, it is instructive to compare the b coefficients in the full model with interactions (Table 7.3b(iii)) with the results of the simple regressions separately computed within each college (see p. 125). The intercepts in the three colleges are \$20,982 (LA), \$28,046 (E), and \$23,600 (BUS). The unweighted mean of the three intercepts is \$24,209.3 which is identical to b_0. The intercept for LA is $-b_1 - b_2 + b_0 = -\$3{,}836.9 - (-\$609.2) + \$24{,}209.3 = \$20{,}982$. Substituting the appropriate values into the equations above for each of the colleges, the intercept for E is $\$3{,}836.9 + \$24{,}209.3 = \$28{,}046$ and the intercept for BUS is $-609.2 + \$24{,}209.3 = \$23{,}600$. Similar logic can be applied to the calculation of the slopes. The slopes in the simple regressions calculated separately for each college were \$790 for LA, \$123 for E, and \$1,872 for BUS. The unweighted mean of the slopes in the three colleges is \$928.3, which equals b_3. The slope for the LA group is $b_3 - b_4 - b_5 = 928.3 - (-805.4) - 943.7 = 790$. The slope for the E group is $b_3 + b_4 = 123$ and the slope for the BUS group is $b_3 + b_5 = 1{,}872$. Thus the differences in the b coefficients between the dummy coding and effect coding analyses directly reflect the differences in meaning. It is important to note that the *simple regression equations* for each group are identical whether dummy coding or unweighted effects coding is used. This is yet another example of the point made in Chapter 3: Predictor scaling does not affect the simple slope analysis for post hoc probing.

Choice of Coding System

Given that the two coding systems discussed above produce results that reflect their different meanings, which coding system should be preferred? When the interactions involve a categorical variable and a continuous variable, dummy variable coding produces immediately interpretable contrasts with the comparison group, whereas simple effect coding does not. Hence, if there is interest in contrasts between pairs of groups, dummy variable coding will be more efficient. When the interactions of interest involve two (or more) categorical variables, effect coding is preferred because it produces results that are immediately comparable with standard ANOVA procedures. For example, when there are equal *n*s in each cell, effect coding produces main effects and interactions that are orthogonal just as in ANOVA. However, dummy coding produces correlations between the contrast vectors for the main effects and those for the interactions. Thus some (minor) adjustments are needed in the results of the dummy coded analysis to produce orthogonal estimates of the variance resulting from the main effects and interactions (see Pedhazur, 1982, p. 369).

Centering Revisited

After the emphasis on centering predictor variables in cases of interactions between two or more continuous variables, the failure to use centered dummy or effect variables (i.e., mean = 0) is striking. However, with categorical variables we are nearly always interested in regression of the predictor variable within the distinct groups themselves rather than at the value of the (weighted) mean of the groups. As we have seen, both dummy coding and effects coding lead to clearly interpretable results in terms of the slopes and intercepts for each group. If, however, we are interested in the *average effect* of the continuous predictor variable, another coding system, weighted effects coding, should be used. Weighted effects codes follow the same logic as the unweighted codes, but take each group's sample size into account. Darlington (1990) presents a discussion of weighted effects codes. Note that in the special case where the sample sizes in each group are equal, unweighted and weighted effects codes are equivalent.

Post Hoc Probing of Significant Interactions

The significant overall (joint) test of the interaction of a categorical and continuous variable tells us only that there is an overall difference in the slopes of the regression lines. As was the case for a significant interaction between two continuous variables, we now wish to probe further the interaction to assist in its interpretation. Three sets of tests that address different questions can be performed.

First, we may test the simple slopes for the regression of the continuous variable (e.g., GPA) on the outcome variable (e.g., starting salary). Because one of the predictor variables is categorical, the simple slopes of interest will be those evaluated at values of the dummy (or effect) variables that correspond to the separate groups. Thus, in our example, we will be interested in evaluating whether the simple slopes for the LA, E, and BUS groups each differ from zero.

Second, we may test the difference between the predicted values in any pair of groups for a specific value of the continuous variable. For example, we may wish to test whether the E and BUS regression lines differ for students who have a specific value of GPA, say 3.5, corresponding to the cutoff for Dean's list.

Third, we may be interested in identifying the region(s) of the continuous variable where two regression lines can be shown to differ significantly. For example, for what range of values of GPA do the E and BUS students differ in their starting salaries?

Testing Simple Slopes Within Groups

To test the significance of the simple slopes within each level of the categorical variable (groups), the general procedures developed in Chapter 2 are followed. For our starting salary example, b_3 is the simple slope for the LA group, $(b_3 + b_4)$ is the simple slope for the E group, and $(b_3 + b_5)$ is the simple slope for the BUS group. The corresponding standard errors for the three groups are $(s_{33})^{1/2}$, $(s_{33} + s_{44} + 2s_{34})^{1/2}$, and $(s_{33} + s_{55} + 2s_{35})^{1/2}$, respectively, where s_{33}, s_{44}, s_{55}, s_{34}, and s_{35} are taken from $\mathbf{S}_b$, the variance–covariance matrix of the predictors.[5] The t-tests can be computed by dividing the simple slope by its corresponding standard error, with df $= 1, n - k - 1$, where k is the number of terms in the regression equation not including the intercept (here $k = 5$).

Computer Procedure

A very simple computer procedure can be used to test the simple slopes in each of the groups. In our example, when $D_1 = 0$ and $D_2 = 0$, the test of the b_3 coefficient in the overall analysis including the categorical variable, the continuous variable, and their interaction (see Table 7.3a(iii)) provides the proper test of the simple slope in the comparison (LA) group. We can take advantage of this fact by noting that the simple slope of the *comparison group* in the particular dummy coding system is always properly tested in this case. In our example, if we recode the groups according to the dummy coding procedure shown in Table 7.1b, the E group is now the comparison group; its simple slope is $b_3 = 122.9$ and the test of b_3 in the regression analysis ($t = 0.65$, ns) provides the appropriate test of the simple slope. Similarly, if we recode the groups accourding to the dummy coding procedure shown in Table 7.1c, $b_3 = 1{,}872.0$ is now the simple slope of the BUS group and the test of b_3 ($t = 11.29$, $p < .001$) provides the appropriate test of significance. Thus, in our case involving three groups, conducting three separate regression runs in which each group in turn serves as the comparison group produces proper tests of each of the three simple slopes.

Differences Between Regression Lines at a Specific Point

A second way to probe significant interactions is to test whether the predicted values for any pair of the groups differ at a specified value of the continuous variable. The Johnson–Neyman technique (Johnson & Fay, 1950; Johnson & Neyman, 1936; Rogosa, 1980, 1981; see Huitema, 1980, for an excellent presentation of the basic technique and its extensions to more complex problems) has long offered a solution to this question. Although the calculations using this technique are straightforward, they are fairly tedious; computer software to perform this analysis is not available to our knowledge in the major statistical packages. We offer here a simple alternative to the Johnson–Neyman technique that draws on our computer method for testing simple slopes.

Let us simplify our example and consider only two groups, E and BUS. Suppose we are interested in determining whether there is a significant difference between the two regression lines at the point where GPA = 3.5 (Dean's list). We will arbitrarily call BUS the comparison group so that our dummy variable $D_1 = 1$ for E and 0 for BUS. The following regression equation describes the two-group case with a potential interaction:

$$\hat{Y} = b_1 D_1 + b_2 \text{ GPA} + b_3 (D_1 \times \text{GPA}) + b_0$$

We can estimate this equation using several different transformations of GPA obtaining different values for the lower order coefficients while the b_3 coefficient remains constant. Below we report three solutions: (a) untransformed GPA on the usual 0.0–4.0 scale; (b) GPA-C centered for the entire sample of 50 students (GPA − 2.78) to be maximally comparable with the centered analyses reported above; and (c) transformed GPA-D = GPA − 3.50, which sets the *0.0* value for transformed GPA equal to the untransformed value of 3.50, the Dean's list cutoff. The results of the three equations are

$$\text{(a) GPA:} \quad \hat{Y} = 9{,}303.4(D_1) + 1{,}872.0(\text{GPA}) + (-1{,}749.1)(D_1 \times \text{GPA}) + 18{,}401.5$$

$$\text{(b) GPA-C:} \quad \hat{Y} = 4{,}446.1(D_1) + 1{,}872.0(\text{GPA-C}) + (-1{,}749.1)(D_1 \times \text{GPA-C}) + 23{,}600.1$$

(c) GPA-D: $\hat{Y} = 3{,}185.5(D_1) + 1{,}872.0(\text{GPA-D})$

$$+ (-1{,}749.1)(D_1 \times \text{GPA-D}) + 24{,}953.5$$

Note that in each case, the b_1 coefficient represents the distance between the regression lines for E versus BUS when the value of the GPA variable is 0.0. For (a), this distance is evaluated when original GPA is 0.0, which is likely to be of little usefulness. For (b), this distance is evaluated when original GPA is 2.78, which is the mean of the entire sample of students. For (c), this distance is evaluated when original GPA is 3.50, which corresponds to our point of interest, the Dean's list cutoff. Thus we find that at GPA = 3.5 the difference in the predicted starting salaries of E and BUS students is $3,186. The test of b_1 is also reported in standard regression packages, $t = 12.01$, $p < .001$, and corresponds to the results of the Johnson–Neyman test of significance of the difference between two regression lines at a GPA of 3.5. This computer solution can also be extended to more complex problems such as testing the significance of the distance between two regression planes or two regression curves when specific values are given for each of the continuous predictor variables.

Several comments about this computer test should be noted.

1. The test should be used with dummy codes because the interest is in comparing differences between group regression lines. Recall that unweighted effects codes compare the regression line with the unweighted mean.

2. When more than two groups are employed, the test of the coefficient for each dummy code provides a test of the difference between the regression lines for the comparison group and the group specified by the dummy code. Note that these tests use the mean square residual (MS_{res}) from the overall regression analysis based on all groups rather than just the MS_{res} from the two groups used in the contrast.

3. Contrasts not involving the designated comparison group can be performed by respecifying the dummy coding as we earlier illustrated for tests of simple slopes.

4. When several pairs of regression lines are being compared, researchers may wish to use the Bonferoni procedure to adjust their obtained values for the number of different tests that are undertaken. Huitema (1980) presents an extensive discussion of the use of the Bonferoni procedure in this context.

Identifying Regions of Significance

Potthoff (1964) has extended the basic Johnson–Neyman procedure to identify *regions* in which the two regression lines are significantly different for all possible points. The goal of the test is to identify regions such that "in the long run, not more than 5% of such regions which are calculated will contain any points at all for which the two groups are equal in expected criterion score" (Potthoff, 1964, p. 244). The test thus allows for the fact that there will be variability from sample to sample in the estimates of the slopes and intercepts of the two regression lines, so that the point at which they begin to differ will also vary concomitantly. The traditional Johnson–Neyman test provides an appropriate test of the difference between two regression lines at a particular value. The equation for the Potthoff test closely follows that of the original Johnson–Neyman test, except that $2F_{2,\,N-4}$ replaces $F_{1,\,N-4}$ as the critical F-value. Discussions and derivations of the two procedures can be found in Potthoff (1964) and Rogosa (1980, 1981). Below we illustrate this analysis by comparing the E and BUS groups from our example.

To perform the analysis, three separate steps are performed following the layout recommended by Pedhazur (1982). First, two separate regression analyses are performed. Second, the results of the regression analyses are used to calculate three intermediate quantities. Third, the cutoff values for the regions of significance are calculated.

Step I: Regression Analyses

For ease of interpretation, the two following regression analyses should be run using the original values (i.e., not centered) of the continuous predictor variable.

1. The continuous predictor variable is regressed on the outcome variable using only the data from group 1:

$$\text{For E only:} \quad \hat{Y}_E = b_{1(\text{E})}(\text{GPA}) + b_{0(\text{E})}$$

2. The continuous predictor variable is regressed on the outcome variable using only the data from group 2:

$$\text{For BUS only:} \quad \hat{Y}_{(\text{BUS})} = b_{1(\text{BUS})}(\text{GPA}) + b_{0(\text{BUS})}$$

Step II: Calculation of Intermediate Quantities

Three intermediate quantities, which are conventionally labeled A, B, and C, must then be calculated. The variables involved in the equations are defined immediately below the formulas.

$$A = \frac{-2F_{2,\,N-4}}{N-4}(\mathrm{SS}_{\mathrm{res}})\left[\frac{1}{\mathrm{SS}_{X(1)}} + \frac{1}{\mathrm{SS}_{X(2)}}\right] + [b_{1(1)} - b_{1(2)}]^2$$

$$B = \frac{2F_{2,\,N-4}}{N-4}(\mathrm{SS}_{\mathrm{res}})\left[\frac{\overline{X}_1}{\mathrm{SS}_{X(1)}} + \frac{\overline{X}_2}{\mathrm{SS}_{X(2)}}\right] + [b_{0(1)} - b_{0(2)}][b_{1(1)} - b_{1(2)}]$$

$$C = \frac{-2F_{2,\,N-4}}{N-4}(\mathrm{SS}_{\mathrm{res}})\left[\frac{N}{n_1 n_2} + \frac{\overline{X}_1^2}{\mathrm{SS}_{X(1)}} + \frac{\overline{X}_2^2}{\mathrm{SS}_{X(2)}}\right] + [b_{0(1)} - b_{0(2)}]^2$$

$F_{2,\,N-4}$ is the tabled F-value with 2 and $N - 4$ degrees of freedom.

N is the number of subjects in the entire sample. n_1 is the number of subjects in group 1; n_2 is the number of subjects in group 2. $N = n_1 + n_2$.

$\mathrm{SS}_{\mathrm{res}}$ is the total residual sum of squares. It is computed by summing the residual sum of squares from the two within-group regression analyses (i.e., regression analysis 1 and regression analysis 2).

$\mathrm{SS}_{X(1)}$ is the sum of squares associated with the predictor variable in group 1 (regression analysis 1); $\mathrm{SS}_{X(2)}$ is the sum of squares associated with the predictor variable in group 2 (regression analysis 2).

$\overline{X}_1$ is the mean of group 1; $\overline{X}_2$ is the mean of group 2.

$b_{1(1)}$ is the slope for group 1 (analysis 1); $b_{1(2)}$ is the slope for group 2. $b_{0(1)}$ is the intercept for group 1 (analysis 1); $b_{0(2)}$ is the intercept for group 2.

Step III. Calculation of Region Cutoffs

The final step is the calculation of the cutoff values for the regions of significance, the formula for which is given below.

$$X = \frac{-B + (B^2 - AC)^{1/2}}{A} \quad \text{and} \quad X = \frac{-B - (B^2 - AC)^{1/2}}{A}$$

Note that these equations do *not* always yield two solutions within the effective range of the predictor variable. Depending on the nature of the interaction, there may be 0, 1, or 2 regions within the possible range of the predictor variable in which the predicted values of the two regression lines differ.

We will illustrate the calculation of the regions of significance for the two-group case using just the data from the E and BUS groups in our example (i.e., we assume E and BUS comprise the entire sample). The necessary data and their source for the computations are presented below.

$$n_1 = 10; \quad n_2 = 15; \quad N = n_1 + n_2 = 25 \text{ (from Table 7.2)}$$

$$F_{2,\,N-4} = F_{2,\,21} = 3.47 \text{ for alpha } = .05 \text{ (from } F\text{-Table)}$$

From Regression Analysis 1 for E group

$$\overline{\text{GPA}}_{(1)} = 2.40; \quad b_{1(1)} = 122.9; \quad b_{0(1)} = 27{,}705.0;$$

$$\text{SS}_{X(1)} = 21{,}768.4; \quad \text{S}_{\text{res}(1)} = 357{,}394.5$$

From Regression Analysis 2 for BUS group

$$\overline{\text{GPA}}_{(2)} = 2.99; \quad b_{1(2)} = 1{,}872.0; \quad b_{0(2)} = 18{,}401.6;$$

$$\text{SS}_{X(2)} = 6{,}671{,}180.4; \quad \text{SS}_{\text{res}(2)} = 613{,}528.5.$$

From Regression Analyses 1 and 2 combining the groups

$$\text{SS}_{\text{res}} = \text{SS}_{\text{res}(1)} + \text{SS}_{\text{res}(2)} = 970{,}923$$

When these values are substituted into the formulas above, and steps II and III are carried out, the values 5.19 and 5.45 are obtained. Recall that we calculated earlier in Chapter 7 (p. 125) that the regression lines for the E and BUS groups had a crossing point of 5.31. Thus, for values of GPA less than 5.19, the E group is predicted to have higher starting salaries than the BUS group, for values of GPA greater than 5.45, the BUS group is predicted to have higher starting salaries than the E group, and for values of GPA between 5.19 and 5.45, the starting salaries of the two groups are not predicted to differ. Given that the possible range of GPA is from 0.0 to 4.0, this means that the E group will always be predicted to have a higher starting salary than the BUS group in the possible range of GPA.

A few final observations should be made about the Potthoff procedure for determining regions of significance.

1. The calculations of steps II and III are quite tedious and are currently unavailable in most standard computer packages. Appendix C of this book contains a simple SAS program for Potthoff's extension of the Johnson–Neyman procedure for the simple two-group case. Borich (1971; Borich & Wunderlich, 1973) offers more extensive computer programs for the Johnson–Neyman procedure.

2. Even when regions of significance are obtained within the *possible* range of the predictor variable, caution should be taken in interpretation. If few or no data points actually fall in the regions, the result represents a serious extrapolation beyond the available data, raising concerns about the meaning of the obtained region. For example, a region of significance less than a cumulative GPA of 1.0 would not be particularly meaningful because few, if any, students ever graduate with such low GPAs. Finally, if the test does not identify any regions of significance within the range of the predictor variable, this indicates that the two regression lines differ for either (a) all values or (b) no values of the predictor variable. In case (a), the b_1 coefficient for the group effect will typically be significant, whereas in case (b), it will not be significant.

3. When regions are being calculated for several pairs of regression lines, researchers may wish to substitute the more conservative Bonferoni F for the F-value listed to maintain the studywise error rate at the level claimed (e.g., alpha = .05).

4. Huitema (1980) includes an extensive discussion of applications of the basic Johnson–Neyman procedure to more complex situations. Note, however, that he presents the test that is appropriate when a priori values of the predictor variables have been selected. Once again, the Potthoff extension requires that $2F_{2,\,N-4}$ be used in place of $F_{1,\,N-4}$ as the critical F-value in the equations.

5. Cronbach and Snow (1977) raise the important methodological point that the interpretation of the regions of significance is clearest in experimental settings in which subjects are randomly assigned to treatment groups. This practice eliminates the possibility that specification error in the regression equation biases the results. Cronbach and Snow present an excellent discussion of the design and analysis of research on aptitude × treatment interactions.

Summary

In this chapter we have discussed the treatment of interactions involving categorical and continuous variables. We have discussed the interpretation of the coefficients in a series of regression equations of increasing complexity using two coding systems for the categorical variable, dummy coding and simple effect coding. We have also discussed post hoc probing of significant interactions including testing of simple slopes, differences between predicted values for pairs of groups, and determination of critical regions of significance. As we have seen, the procedures developed in previous chapters for the interpretation of interactions between two or more continuous predictor variables generalize nicely to interactions involving categorical and continuous variables. Finally, although not discussed in the present chapter, the prescriptions developed in Chapter 6 for model and effect testing with higher order terms are directly applicable to regression equations containing categorical predictor variables.

Notes

1. In the dummy coding system of Table 7.1a, dummy code D_1 actually contrasts E to LA and E to BUS. Dummy code D_2 actually contrasts BUS to E and BUS to LA. The two contrasts share in common the E versus BUS contrast. When D_1 and D_2 are entered into the same regression equation, the b_1 regression coefficient (for D_1) reflects that part of the D_1 contrast that is independent of D_2, that is, the E versus LA contrast. The b_2 coefficient (for D_2) reflects that part of the D_2 contrast that is independent of D_1, that is, the BUS versus LA contrast.

2. The simplest method of directly comparing the E and BUS groups is to rerun the regression analysis using a dummy coding system that designates one of these groups as the comparison group. In Table 7.1, both sections (b) and (c) include this comparison. This method can be used in the more complex models described below and is illustrated later in the chapter.

3. If we had not centered GPA, b_0 would have represented the predicted value of starting salary for LA graduates with a GPA of 0.0. Presumably, none of these individuals would actually graduate, so that this predicted salary value would not be meaningful.

4. In previous instances of rescaling of first order terms by additive constants, there has been no resultant change in the coefficient for the interaction. The reader should be aware that the change from dummy codes to effect codes is not a mere change in scaling by additive constants. The interaction coefficients do change with a change in coding scheme.

5. The weight vector used to compute the standard errors is $w' = [0 \quad 0 \quad 1 \quad D_1 \quad D_2]$, where the values of D_1 and D_2 are (0 0) for LA, (1 0) for E, and (0 1) for BUS. The general expression for the standard error of the simple slopes is $w'\mathbf{S}_b w$, where $\mathbf{S}_b$ is the 5×5 variance-covariance matrix of the b coefficients available from standard computer packages.

8 Reliability and Statistical Power

Throughout the previous chapters we have made no mention of measurement error in the predictors in regression analysis. We have assumed a regression model in which predictors are measured without error. In simple linear regression analysis with no higher order terms, measurement error in predictors introduces bias into the estimates of regression coefficients (i.e., the expected values of the regression coefficient estimates no longer equal the population parameters). The same is true in more complex models. Measurement error in individual predictors produces a dramatic reduction in the reliability of the higher order terms constructed from them. In turn, reduced reliability of higher order terms increases their standard errors and consequently reduces the power of their statistical tests.

Moreover, the power of statistical tests for higher order terms, even in the absence of measurement error, is expected to be low. Most recently, these concerns have been a particular focus in the literature on moderated multiple regression (Chaplin, in press-a), which addresses tests of models involving interactions. Recent discussions of issues of power and reliability in moderated multiple regression are given in Arvey, Maxwell, and Abraham (1985), Champoux and Peters (1987), Cronbach (1987), Dunlap and Kemery (1987, 1988), Evans (1985), Lubinski and Humphreys (1990), Paunonen and Jackson (1988). The concerns expressed in this literature and the findings have implications for all equations considered in this text.

In this chapter we first review classic concepts of reliability and introduce an analysis of the reliability of interaction terms. We consider strategies that have been proposed to correct for the unreliability of product terms in regression analysis. We also consider the issue of whether spurious effects can be produced by measurement error. Then we turn to the assessment of the statistical power of tests of interactions. We consider first the number of cases required to achieve adequate power in the case of no measurement error in predictors. Then we consider the impact of measurement error on effect sizes, statistical power, and sample size requirements.

The first section of the chapter, Reliability, presents the theoretical basis for the effects of measurement error on regression estimates. Once again, readers may find this section to be slower going than the remainder of the book. The second section, Statistical Power, shows the actual impact of measurement error on statistical power and sample size requirements for tests of the interaction. This section can be read without a full understanding of the theoretical material in the first section.

Reliability

Biased Regression Coefficients with Measurement Error

Our discussion of measurement error will initially focus only on the reliability of the predictor variables, because measurement error in the criterion does not introduce bias into unstandardized regression coefficients (see, e.g., Duncan, 1975). We begin with a brief review of the definition of reliability and consideration of the effect of unreliability of a predictor in the simple one-predictor regression analysis (see Bohrnstedt, 1983; Duncan, 1975; Heise, 1975; Kenny, 1979).

The Single Predictor Case

In classical measurement theory (Gulliksen, 1987), an observed score (X) is defined as being comprised of the true score on the variable (T_X) + random error (ϵ_X), that is, $X = T_X + \epsilon_X$. This definition implies a linear relationship between X and T_X. Classical measurement theory assumes that:

1. The mean or expected value of the random error in the population is zero, that is, $E(\epsilon_X) = 0$;
2. that the errors are normally distributed;

3. that the covariance between the random errors and true scores is zero, that is, $C(T_X, \epsilon_X) = 0$.

From these assumptions, it follows that the variance of observed scores is comprised of two components, true score variance and error variance as follows:

$$\sigma_X^2 = \sigma_{T_X}^2 + \sigma_{\epsilon_X}^2 \tag{8.1}$$

Then the reliability of the variable is defined as the proportion of total variance in X that is true score variance.

$$\rho_{XX} = \sigma_{T_X}^2/\sigma_X^2 = \sigma_{T_X}^2/[\sigma_{T_X}^2 + \sigma_{\epsilon_X}^2] \tag{8.2}$$

Bias Due to Unreliability in the Regression Coefficient and Correlations. How does measurement error in the predictor X and the criterion Y affect the covariance between them? Given $X = T_X + \epsilon_X$ and $Y = T_Y + \epsilon_Y$, the covariance between X and Y is given as follows:

$$C(X, Y) = C(T_X, T_Y) + C(T_Y, \epsilon_X) + C(T_X, \epsilon_Y) + C(\epsilon_X, \epsilon_Y) \tag{8.3}$$

However, this expression may be simplified with two further assumptions of classical measurement theory: Random errors of the predictor and criterion are uncorrelated with true scores, so that $C(T_X, \epsilon_Y) = C(T_Y, \epsilon_X) = 0$, and the important assumption that all random errors are uncorrelated, that is, $E(\epsilon_X \epsilon_Y) = 0$. Under these assumptions the observed covariance equals the true covariance:

$$C(X, Y) = C(T_X, T_Y) \tag{8.4}$$

Under classical measurement theory, covariances are unaffected by measurement error. The unstandardized coefficient for the regression of Y on X is given as

$$b_{YX} = C(X, Y)/\sigma_X^2 = C(T_X, T_Z)/[\sigma_{T_X}^2 + \sigma_{\epsilon_X}^2] \tag{8.5}$$

Thus, if there is measurement error in predictor X, then b_{YX} is biased: b_{YX} will be closer to zero than the population value it estimates. This bias occurs because the denominator of b_{YX} is the observed variance of predictor X, which is inflated by measurement error. We will refer to bias in which estimates are closer to zero than corresponding parameters as *at-*

tenuation bias or simply *attenuation* throughout this discussion. Measurement error also produces attenuation in simple (zero order) correlations ($\rho_{XY} = C(X, Y)/\sigma_X\sigma_Y$), because the denominator of the correlation coefficent contains the observed standard deviations of the variables.

Multiple Regression with no Product Terms

Under classical measurement theory assumptions, the variances of the predictors are inflated by measurement error, but the observed covariances are not. This does not mean, however, that measurement error invariably causes attenuation of regression estimates in the multiple predictor case. With several fallible first order predictors, the *extent* and *direction* of bias in each regression coefficent depends upon the intercorrelations among the *true scores* underlying measured variables (see demonstrations by Bohrnstedt & Carter, 1971; Cohen & Cohen, 1983; Kenny, 1979).

Cohen and Cohen (1983) provide a very clear demonstration of this point. In the two-predictor regression of Y on X and Z, the standardized partial regression coefficient for Y on X is given as:

$$b_{YX.Z} = \frac{r_{YX} - r_{YZ}r_{XZ}}{1 - r_{XZ}^2}$$

If variable Z has reliability ρ_{ZZ} less than 1.0, then the numerator of $b_{YX.Z}$, corrected for measurement error, would be $r_{ZZ}r_{YX} - r_{YZ}r_{XZ}$, where r_{ZZ} is the sample estimate of ρ_{ZZ}. The observed value of the numerator of $b_{YX.Z}$, taking into account the unreliability of the partialled predictor Z, can vary widely from the true value. A true nonzero relationship between X and Y in the *population* may not be observed at all in the *sample*; a true zero relationship between X and Y may appear to be nonzero; and even the sign of the observed regression of Y on X may vary from the true value in a particular sample. The reliability of the *partialled* variable (here Z) has a profound effect on the bias in the estimators for other variables. Even if one predictor in a set is perfectly reliable, its regression coefficient is subject to bias produced by error in other predictors. Only in the case in which the *true scores* underlying observed predictors are uncorrelated with one another is each coefficient guaranteed to be attenuated by measurement error, as in the single predictor case (Maddala, 1977).

Regression with Product Terms

Measurement error in product terms (e.g., XZ, X^2Z) that represent interactions in regression requires special consideration. Of particular concern is the covariance between errors of measurement in the product terms and in the predictors of which the product terms are comprised.

Covariance Between Errors in Product Term and Components. Regression models containing product terms have all the problems of unreliability presented thus far. However, product terms also introduce another complexity: Even under the assumptions of classical measurement theory, the covariances between the product term and its components are affected by measurement error (see, e.g., Marsden, 1981).

Consider the following expression for the observed crossproduct between two variables, taken from Bohrnstedt & Marwell (1978, p. 65):

$$XZ = (T_X + \epsilon_X)(T_Z + \epsilon_Z) = T_X T_Z + T_X \epsilon_Z + T_Z \epsilon_X + \epsilon_X \epsilon_Z \quad (8.6)$$

The last three terms $(T_X \epsilon_Z + T_Z \epsilon_X + \epsilon_X \epsilon_Z)$ represent the measurement error component (ϵ_{XZ}) of the observed XZ product. These terms have nonzero covariance with the error in their components. For example:

$$\begin{aligned} C(\epsilon_X, \epsilon_{XZ}) &= C[\epsilon_X, (T_X \epsilon_Z + T_Z \epsilon_X + \epsilon_X \epsilon_Z)] \\ &= T_X C(\epsilon_X, \epsilon_Z) + T_Z \sigma^2_{\epsilon_X} + C(\epsilon^2_X, \epsilon_Z) \end{aligned} \quad (8.7)$$

Under the assumptions of bivariate normality and uncorrelated errors, this expression reduces to

$$C(\epsilon_X, \epsilon_{XZ}) = T_Z \sigma^2_{\epsilon_X} \quad (8.8)$$

so that the observed covariance between a crossproduct term and a component is as follows:

$$C(XZ, X) = C(T_X T_Z, T_X) + T_Z \sigma^2_{\epsilon_X} \quad (8.9)$$

The covariance of errors in XZ and errors in Z will also produce bias in regression coefficients for the crossproduct term and its components. The bias is introduced into the regression coefficients, because $C(XZ, X)$ and $C(XZ, Z)$, both of which contain measurement error, are terms in these coefficients.

Bias in the Regression Coefficient for the Product Term. Once again the direction of bias depends upon the correlation between *true* scores of predictors, with the same vagaries as in multiple regression with no higher order terms. Under the assumption of bivariate normality, the following expression from Bohrnstedt and Marwell (1978) gives the correlation between true scores in XZ and in X:

$$\rho_{T_X T_Z, T_X} = \frac{\dfrac{\mu_X\, C(X, Z) + \mu_Z \sigma_X^2 \rho_{XX}}{(\sigma_{XZ}^2 \sigma_Z^2)^{1/2}}}{\rho_{XZ,XZ}\, \rho_{XX}} \tag{8.10}$$

where $T_X T_Z$ and T_X are the true scores associated with observed variables XZ and X; σ_{XZ}^2 and σ_Z^2 are the variances of XZ and X, respectively; and μ_X and μ_Z are the population means of X and Z, respectively. Inspection of the numerator of equation 8.10 shows that this correlation goes to zero if X and Z are bivariate normal and centered in the population such that $\mu_X = \mu_Z = 0$. For the centered case in the population, when the crossproduct term is less than perfectly reliable, the regression coefficient for that term will be attenuated, just as in the one-predictor case. Only in the centered case in the population with bivariate normal predictors X and Z is the uncertainty of the direction of bias of this term removed; the interaction stands alone as does the single predictor, because the true scores of the interaction term and true scores of its components are uncorrelated, that is, $C(T_X, T_X T_Z) = C(T_Z, T_X T_Z) = 0$.

Reliability of the Crossproduct Term. Bohrnstedt and Marwell (1978) derive an expression for the reliablity of X and Z under the assumptions of classical measurement theory *plus the assumption of bivariate normality* of the predictors:

$$\rho_{XZ,XZ} = \frac{\theta_X^2 \rho_{ZZ} + \theta_Z^2 \rho_{XX} + 2\theta_X \theta_Z \rho_{XZ} + \rho_{XZ}^2 + \rho_{XX} \rho_{ZZ}}{\theta_X^2 + \theta_Z^2 + 2\theta_X \theta_Z \rho_{XZ} + \rho_{XZ}^2 + 1} \tag{8.11}$$

where $\rho_{XZ,XZ}$ is the reliability of the crossproduct term; $\theta_X = \mu_X/\sigma_X$ is the ratio of the mean to the standard deviation of X; $\theta_Z = \mu_Z/\sigma_Z$ is the ratio of the mean to the standard deviation of Z; ρ_{XX}, ρ_{ZZ} are the reliabilities of X and Z; and ρ_{XZ} is the correlation between X and Z.

When X and Z are centered in the population, then $\mu_X = \mu_Z = 0$, and this expression reduces to the simpler form given in Busemeyer and Jones (1983):

$$\rho_{XZ,XZ} = \frac{\rho_{XZ}^2 + \rho_{XX} \rho_{ZZ}}{\rho_{XZ}^2 + 1} \tag{8.12}$$

If X and Z are uncorrelated, then the reliability of the product term is simply the product of the reliability of the components. Bohrnstedt and

Marwell (1978) point out the disconcerting fact that $\rho_{XZ,XZ}$ depends on the scaling of variables. Nevertheless, expression 8.12 is instructive, in that it permits examination of the reliability of the crossproduct XZ in terms of the reliability of its components X and Z, when the predictor variables are centered in the population. Table 8.1 shows the reliability of X and Z required to produce a specified reliability of the crossproduct term, as the correlation between X and Z varies. When X and Z are uncorrelated, they must each have reliabilities of .89 in order that the crossproduct have a reliability of .8. When the individual predictors each have good reliabilities (.84), the reliability of the crossproduct XZ is only .70. Note that as the interpredictor correlation increases, the reliability of the crossproduct term increases slightly; or equivalently, slightly lower reliabilities of the X and Z variables are required to produce a crossproduct term with a specified reliability. What is clear from Table 8.1 is that the individual variables entering a crossproduct term must be highly reliable if even adequate reliability is to be achieved for the crossproduct term.

Corrected Estimates of Regression Coefficients in Equations Containing Higher Order Terms

One approach researchers have taken to the problem of measurement error in regression analysis is to attempt to correct regression coefficients for measurement error. The typical strategy is to correct the covariance or correlation matrix of the predictors for the error, yielding matrices said

Table 8.1
Individual Variable Reliabilities Required To Produce a Specified Crossproduct Reliability, as a Function of the Interpredictor Correlation

Desired Crossproduct Reliability	Interpredictor Correlation ρ_{XZ}			
	0	.10	.30	.50
.90	.95	.95	.94	.94
.80	.89	.89	.88	.87
.70	.84	.83	.82	.79
.60	.77	.77	.75	.71
.50	.71	.70	.67	.61
.40	.63	.63	.59	.50

NOTE: Each entry is the reliability of X and of Z (ρ_{XX} and ρ_{ZZ}) required to produce a specified reliability of their crossproduct $\rho_{XZ,XZ}$. This value depends upon the correlation between X and Z. For example, for a crossproduct reliability of .70, given that ρ_{XZ} = .30, each variable must have a reliability of .82 (alternatively, the product of the reliabilities $\rho_{XX}\rho_{ZZ}$ must be $.82^2$).

to be *corrected for attenuation* (Fuller & Hidiroglou, 1978). These matrices are then used to generate corrected regression coefficients. The approach historically has been most commonly used in path analysis (Kenny, 1979) to correct structural or path coefficients.[1] These coefficients are critical for theory testing, because the sizes of these coefficients are used to infer the magnitude of causal effects of the variables. These approaches require estimates of the reliabilities, which are provided for the first order predictors by work in measurement theory and test construction (Gulliksen, 1987, Lord & Novick, 1968; Nunnally, 1978).[2] Then an expression for the reliability of the crossproduct term can be determined based on its component reliabilities.

A second approach to obtaining corrected estimates of regression coefficients for product terms is through the use of latent variable structural models (Kenny & Judd, 1984). In this approach theoretically error-free latent variables are estimated, which then are used to compute the regression coefficients (structural coefficients).

Correcting the Covariance Matrix: Estimates from Classical Measurement Theory

This approach uses classical measurement theory to attempt to disattenuate the covariance matrix of the predictors (i.e., to eliminate error from the observed variances and covariances). The corrected covariance matrix is then used in the regression analysis (Busemeyer & Jones, 1983).

Variances of First Order Terms. Variances of the first order terms are corrected for error using equation 8.1:

$$\sigma^2_{T_X} = \sigma^2_X - \sigma^2_{\epsilon_X} \tag{8.13}$$

that is, an estimate of the error variance is subtracted from the observed variance of each predictor. Given known reliability ρ_{XX}, the error variance is given as $\sigma^2_{\epsilon_X} = \sigma^2_X(1 - \rho_{XX})$.

Covariances Between First Order Predictors and of First Order Predictors with the Criterion. Under the assumption of classical measurement theory that errors are uncorrelated, covariances between pairs of first order predictors and between first order predictors and the criterion are unaffected by measurement error (see equation 8.4). Hence no correction is required in the covariances among first order terms or of first order terms with the criterion.

Variance of the Product Term. The variance of the product term σ^2_{XZ} is a complex function of the means and variances of the components and the covariance between the components (see Appendix A for a derivation of the following expression):

$$\sigma^2_{XZ} = T^2_X\sigma^2_Z + T^2_Z\sigma^2_X + 2C(X, Z)T_XT_Z + \sigma^2_X\sigma^2_Z + C^2(X, Z) \quad (8.14)$$

A portion of the observed variance of the crossproduct term is due to error. The error is also a complex function of the means and variances of each of the first order components, as well as of their reliabilities. Bohrnstedt and Marwell (1978) provide an expression for error variance contained in the crossproduct term in terms of parameters of the first order terms:

$$\sigma^2_{\epsilon_{XZ}} = T^2_Z\sigma^2_{\epsilon_X} + T^2_X\sigma^2_{\epsilon_Z} + \sigma^2_{\epsilon_X}\sigma^2_Z\rho_{ZZ} + \sigma^2_{\epsilon_Z}\sigma^2_X\rho_{XX} + \sigma^2_{\epsilon_X}\sigma^2_{\epsilon_Z} \quad (8.15)$$

To correct the variance of the crossproduct term for error, observed values are substituted into equation 8.15 to obtain an estimate of the error variance contained in the observed variance. Then the observed variance of the crossproduct term is reduced by subtracting the estimate of the error variance.

Covariances of Product Terms with Components. Covariances of crossproduct terms with their components must also be corrected, because, as we have shown (equation 8.9), the observed covariance between a crossproduct term and a component contains measurement error. Equation 8.8 provides an expression for the error variance contained in the covariance between a crossproduct and its component. An estimate of this error variance is again obtained by substituting observed values for parameters. Then the estimate of the error variance is subtracted from the observed covariance between the crossproduct and a component. Note that there will be separate estimates for correcting $C(XZ, X)$ and $C(XZ, Z)$.

Covariance of Product Terms with Other Variables. If the components X and Z of the XZ crossproduct term and another predictor W are multivariate normal, then the covariance of the product term with the other predictor W need not be corrected. This is because third moments of multivariate normal distributions vanish (see Appendix A, equation A.1).

Cautions. There are substantial cautions concerning regression estimates derived from a covariance matrix corrected as specified above.

1. There is the *strong* assumption that errors are uncorrelated, though there is considerable evidence in the social sciences that errors of measurement are correlated when measures are gathered with a common methodology (e.g., Campbell & Fiske, 1959; West & Finch, in press).

2. If the correction is incomplete, that is, the variances and covariances of some variables measured with error are not corrected, then the partial correction introduces, rather than eliminates, bias in the coefficient estimates (Won, 1982).

3. The covariance matrix created by the corrections may not be positive definite, meaning that there is no solution for the regression coefficients.

4. Regression coefficients derived from disattenuated covariance matrices tend to be overcorrected (Cohen & Cohen, 1983).

5. Regression coeffcients derived from disattenuated covariance matrices cannot be tested for significance, even if the corrections are made using the reliability of each variable in the population. Fortunately, improved methods have recently been developed for correcting regression with multiplicative terms for measurement error (Feucht, 1989; Fuller, 1987; Heise, 1986).

Correcting the Covariance Matrix: Correlated Errors Assumed

Heise (1986) proposed a correction for measurement error in order to improve estimates of regression coefficients, including those for multiplicative terms, but with a cost of increased variance of the estimates. He assumed that errors were not independent of one another, that is, $E(\epsilon_X \epsilon_Z) \neq 0$, while retaining all other assumptions of classical measurement theory. Permitting correlated measurement errors produces a divergent result from that of classical measurement theory: The covariances among pairs of first order terms are now affected by error. Heise provided expressions for the error structure of product terms up through products of six variables in terms of sums of squares and crossproducts of observed scores, plus error variances and covariances. This work extends the algebraic developments of Bohrnstedt and Marwell (1978) for the two variable case. All the expressions are given in terms of sums of squares and crossproducts of *observed* scores, plus error variances and covariances. Two expressions (8.16 and 8.17) are provided here to show the effect of the assumption of correlated errors (we have modified Heise's notation to be consistent with our prior usage). For example, the covariance between the two first order variables *X* and *Z* is given as follows:

$$C(X, Z) = C(T_X, T_Z) + C(\epsilon_X, \epsilon_Z) \qquad (8.16)$$

where $C(\epsilon_X, \epsilon_Z)$ is the covariance between the measurement errors in *X* and *Z*. Note that under classical measurement theory, covariances between first order terms were unaffected by measurement error.

The covariance of a crossproduct term with a component is given as follows:

$$C(XZ, X) = C(T_X T_Z, T_X) + T_Z \sigma^2_{\epsilon_Z} + 2T_X C(\epsilon_X, \epsilon_Z) \qquad (8.17)$$

The second and third terms in equation 8.17 represent the error portion of the covariance between the observed crossproduct and the component. A comparison of equations 8.17 and 8.9 shows that permitting errors to be correlated introduces an additional term reflecting the error covariances, namely, $2T_X C(\epsilon_X, \epsilon_Z)$.

To correct the covariance matrix of the predictors for error, Heise substituted sample means for true scores. He estimated error variances and covariances from multiple measurements on each value of each predictor. The mean observation on one such point was taken to represent a true score. The variance of the observations on the single predictor value provided an estimate of the error variance; the covariance between the repeated observations on single cases across two predictors provided an estimate of the covariance between errors. These estimates were pooled across all cases and were used to adjust the covariance matrix of the predictors. Corrections for the variances of second order (*XZ*) and third order (*XZW*) crossproducts and their covariances followed Heise's derived expressions. Reliabilities of the nine individual scales all exceeded .90 (except for one scale for one subgroup of raters). In this case the corrected regression estimates did not differ dramatically from the uncorrected estimates.

In a subsequent simulation study Heise varied sample size ($n = 200$, 350, 500) and the reliability of the first order predictors (.70, .90). Bias was always reduced *on average* with the method, even for the smallest sample size. However, the corrected estimates varied substantially across replications, particularly with reliabilities of .70, and large sample size did not compensate for unreliability.

We are not surprised at the unstable solutions obtained when the reliability of the predictor was .70. If the predictors are uncorrelated, then under the assumption of uncorrelated errors the reliability of the second order crossproduct (*XZ*) term would be $.70^2 = .49$, and the reliability of a third order crossproduct (*XZW*) term, $.70^3 = .34$. No correction method can be expected to salvage analyses containing variables so frought with measurement error. Yet, if our first order predictors have reliabilities of even .70, these are the levels of measurement error in our higher order terms. As a final note, Heise's correction may lead to covariance matrices that are not positive definite; then there is no proper solution for the corrected regression coefficients.[3]

Correcting the Covariance Matrix: Constraint for Proper Solution

The work of Fuller (1980, 1987; Fuller & Hidiroglou, 1978) provides an improvement upon Heise's approach. The corrections applied to the covariance matrix are constrained so that the corrected covariance matrix is positive definite, that is, it has the statistical structure of moment (covariance, correlation) matrices that will yield proper solutions for corrected regression coefficients. The procedure also takes into account the error of estimation in reliabilities when population values of reliability coefficients are unknown. The procedure produces efficient estimates of regression coefficients and consistent estimates of the corrected covariance matrix of the regression coefficients, from which corrected standard errors of the regression coefficients are derived.

Feucht (1989) provides a Monte Carlo simulation study comparing the results of Heise's *corrected estimator* approach, Fuller's *corrected/constrained estimator* approach, and the standard ordinary least squares (OLS) regression approach (without correction) in small samples ($n = 60, 90$) with varying predictor reliability (all .90, all .60, or a mixture of .60, .90). The regression equation examined contained three first order terms and one two-way interaction.

A problem known as matrix indefiniteness (see note 3) of the covariance matrix of predictors resulted from the *corrected estimator* procedure. The prevalence of the problem increased as sample size and reliability decreased, to 54% of all samples with the lowest reliability (.60 for the three first order predictors) and smaller sample size ($n = 60$). Matrix indefiniteness produces improper regression estimates. With regard to bias of estimates, the *corrected/constrained approach* produced less biased estimates, and the *corrected approach* more biased estimates relative to OLS, particularly for the crossproduct term.

Corrected/constrained estimates showed slight attenuation bias, providing a conservative correction, rather than the overadjustment of other methods. In terms of the efficiency of the estimates, OLS produced the most efficient estimates (smallest variance of the estimates across samples), with the *corrected/constrained approach* being less efficient. In contrast, Heise's *corrected approach* produced very large variances, consistent with the findings of Heise's (1986) simulation. In general, performance of all estimators, in terms of both bias and efficiency, deteriorated dramatically with decreases in reliability. Interestingly, mixed reliabilities in the predictors in some cases led to worse performance than uniformly low reliability. Reflecting the importance of sample size, an increase from $n = 60$ to $n = 90$ produced marked improvements in bias and efficiency of all estimates. Corrected/constrained estimators were the best of the three alternative approaches with small samples and low reliability.

Latent Variable Approaches to Estimates of Regression Coefficients in the Presence of Measurement Error

An important alternative approach to the correction of estimates for measurement error is provided by structural equation models with latent variables. These models are the focus of a large and complex literature (see e.g., Long, 1983a, 1983b, for an introduction; Bollen, 1989; Hayduk, 1987; Jöreskog & Sörbom, 1979, for more advanced treatments) and require specialized statistical software to provide estimates and standard errors of parameters. Here, we will only outline the basic procedure and assumptions of the approach for equations containing curvilinear effects or interactions.

The structural equation modeling approach conceptually develops two sets of equations: (a) measurement equations that describe the relationship of each of the measured variables to an underlying latent variable; (b) structural equations that describe the relationship of each of the latent predictor variables to the criterion variable(s). The structural coefficients obtained from (b) correspond to regression coefficients that have been corrected for measurement error.

To understand the measurement portion of the model, imagine a researcher is attempting to measure socioeconomic status (SES). SES is a latent variable that cannot be measured directly, so typically three indicators of the latent variable are measured: income (I), occupation (O), and education (E). Each of these single indicators is an imperfect measure of SES, that is, they contain measurement error. However, the common variance shared by all three indicators provides an excellent representation of the SES latent variable. A measurement model can be constructed to represent SES. The set of measurement equations would be as follows:

$$I = \lambda_1(\text{SES}) + \epsilon_1$$

$$O = \lambda_2(\text{SES}) + \epsilon_2$$

$$E = \lambda_3(\text{SES}) + \epsilon_3$$

In these equations, the λs represent factor loadings, the correlation between the measured variable and the latent variable; the ϵs represent measurement error. The factor loadings and the errors are estimated using specialized software such as LISREL 7 (Jöreskog & Sörbom, 1989) or EQS (Bentler, 1989). With four or more indicators of the underlying latent factor, tests may be performed to determine if a single factor is sufficient to account for the data. The factors produced through this approach are corrected for measurement error.

These factors are then used in the structural part of the model. Conceptually, a regression analysis is performed using the factor scores as predictors and the criterion providing theoretically error-free estimates of the structural (regression) coefficients. In practice, the measurement and structural portions of the model are estimated simultaneously. The estimation technique assumes multivariate normality, that the latent variables are not correlated with the errors, and typically that errors of measurement are uncorrelated. This last assumption can be relaxed through the specification of a precise pattern of correlated errors in some models.

Some work in this area has addressed interactions and curvilinear effects in the structural portion of the model. Alwin and Jackson (1981), and Byrne, Shavelson, and Muthén (1989), and Jöreskog (1971) have considered methods of comparing models across groups, a case corresponding to the interactions between a categorical (group) and continuous (factor) variable considered in Chapter 7. The approach involves comparison of two models: In model 1, the unstandardized structural (regression) coefficients are set to be equal in each group (no interaction); in model 2, the structural coefficients are estimated separately in each group. A comparison of the goodness of fit of the two models provides a test of the interaction term. This approach also permits tests of the equivalence of the measurement portion of the model across the groups. This approach is easily implemented, and several examples exist in the literature (see, e.g., West, Sandler, Pillow, Baca, & Gersten, 1991, for an empirical illustration).

Kenny and Judd (1984) have developed methods for testing interactions and curvilinear effects involving continuous latent variables. They show that, by forming products of the indicator variables, all of the information is available that is needed to estimate models containing X^{*2} and X^*Z^*, where X^* and Z^* are latent variables. For example, to estimate X^{*2} in the curvilinear case with two measured variables, all information needed to estimate the variance and covariance terms in the model can be derived based on the products of the measured variables X_1 and X_2 (i.e., X_1^2, X_2^2, and X_1X_2). For the latent variable X^*Z^* interaction with two measures each of X^* and Z^*, the crossproducts of the measured variables X_1, X_2 and Z_1, Z_2 (i.e., X_1Z_1, X_1Z_2, X_2Z_1, and X_2Z_2) provide the starting point for the estimation of the model. In each case, the products of the measured variables become the indicators for the corresponding latent variable. In a simulation of the performance of the model in the presence of measurement error in the observed variables, Kenny and Judd showed that their approach provided good estimates of the parameters in a known underlying true model.

Bollen (1989) noted three obstacles to the use of the Kenny and Judd approach. First, there was initially considerable difficulty in implementing the procedure in the widely available EQS and LISREL programs, although Hayduk (1987) and Wong and Long (1987) have recently described successful methods. Second, the formation of the products of the indicator variables violates the assumption of multivariate normality that is necessary for the estimation procedure to produce correct standard errors. Alternative estimation procedures (Browne, 1984) exist in the EQS and LISREL programs that do not make this assumption; however, these procedures require large sample sizes to produce proper estimates (Bentler & Chou, 1988; West & Finch, in press). Third, Kenny and Judd assumed that the latent variables and disturbances (term representing unexplained variation in a latent criterion variable) of the components of the latent variables are normally distributed. These assumptions can be tested using the EQS program. Again, violation of this assumption may require the use of alternative estimation procedures or respecification of the model to achieve proper estimates.

The Kenny and Judd approach to correction of measurement error in regression equations involving curvilinear effects or interactions has shown considerable promise to date in a small number of studies. However, the approach has to date been difficult for most researchers to implement, precluding its more frequent use in the literature.

Can Measurement Error Produce Spurious Effects?

We have focused on the extent to which measurement error attenuates regression estimates. It is also possible that measurement error might lead to effects being observed in the sample that do not actually exist in the population, that is, *spurious effects*.

Spurious First Order Effects? In equations containing interactions, first order effects become biased and unstable when predictors contain measurement error. We should not be surprised at this conclusion, given the low reliability of the product term relative to the first order effects. Cohen and Cohen (1983) point out that it is the reliability of the partialled variable (in this context XZ) that has a profound effect on estimates of *other* variables (here X and Z). Evans (1985) found substantial lability in the joint variance accounted for by first order effects in regression equations containing interactions when there was measurement error in the predictors. Feucht's (1989) simulations also showed that in the regression equation $\hat{Y} = b_1X + b_2Z + b_3W + b_4XZ + b_0$, the variances of the estimates of b_1 and b_2 from each of three approaches (OLS, corrected, and corrected/constrained) exceeded those of b_3. Low reliability had a more deleterious effect on both bias and stability of all three estimates of b_1 and b_2 than on the estimates of b_3. Finally, Dunlap and Kemery (1988) found inflated Type I error estimates for the first order coefficients of components of the interaction. These findings suggest that estimates of conditional effects of first order terms in equations containing interactions (see Chapters 3 and 5) are likely to be positively biased in the presence of measurement error. That is, the estimates of the first order terms will be further from zero than their corresponding population values.

Spurious Interactions? Although spurious interactions have not been shown to occur when there are random errors of measurment in the predictors, the effects of other types of measurement error on interactions should also be considered. Evans (1985) investigated the effects of correlated (systematic) measurement error between the predictors and the criterion on estimates of interactions. Correlated measurement error may be expected when similar methods are used to collect measures on the predictor and criterion variables (e.g., all measures are self-report questionnaires). Evans conducted Monte Carlo simulations that varied (a) the magnitude of the interactions in the regression equation (no interactions, weak interactions, and strong interactions), (b) the level of correlated error between predictors and criteria, and (c) the level of predictor reliability (random measurement error). In his study using a large sample size (n = 760) there was no evidence that correlated measurement errors produced spurious interactions, although these systematic errors did attenuate the size of the estimates of interaction effects. Random measurement error attenuated both first order and interactive effects. When the reliability of the predictor variables comprising the interaction was .80, the variance explained by the interaction effect was reduced by half relative to the true variance in the population without measurement error.

Busemeyer and Jones (1983) and Darlington (1990) have noted one set of conditions in which spurious interactions can be produced in observed data even though none exist in the population. The conditions involve *nonlinear* measurement models, for example, $X = k(T_X)^{1/2} + \epsilon_X$, that violate the fundamental assumption of classical test theory that observed scores must be linearly related to the underlying true scores. Thus measurement instruments that have only ordinal rather than interval level properties can produce spurious estimates of interactions and curvilinear effects. Advanced methods of estimating nonlinear measurement models do exist. The interested reader may consult Etezadi-Amoli & McDonald (1983) and Mooijaart & Bentler (1986) for discussion of one class of methods for estimating nonlinear measurement models.

Comment

As anticipated by Cohen and Cohen (1983), the statistical and social science literatures have increasingly addressed the effects of measurement error in regression analysis. More will be learned about the effects of measurement error and behavior of corrected estimates as future simulation studies build on the basic work of Dunlap & Kemery (1987), Evans (1985), and Feucht (1989). New theoretical and empirical developments should continue and the approaches to correcting estimators for measurement error should become more accessible to researchers. Nonetheless, as better and more accessible methods of deriving corrected regression estimates in the presence of measurement error become available, this does not mean that social scientists can relax their concerns about improving measuring instruments. Each of the approaches to correcting for measurement error is based on strong assumptions. If these assumptions are seriously violated, then the corrected regression coefficients will be seriously biased. Highly reliable measures in studies with adequate sample sizes produce the strongest social science.

Statistical Power

Many authors have commented on the weak power of tests for interaction terms in MR, particularly in the face of measurement error (e.g., Busemeyer and Jones, 1983; Dunlap & Kemery, 1988; Evans, 1985). The question has been raised with respect both to interactions involving two (or more) continuous variables as well as to interactions involving categorical and continuous variables (see Chaplin, 1991; in press; Cronbach, 1987; Cronbach & Snow, 1977; Dunlap & Kemery, 1987; Morris, Sherman, & Mansfield, 1986; Stone & Hollenbeck, 1989). In this section we explore the power of tests of the interaction in the equation $\hat{Y} = b_1X + b_2Z + b_3XZ + b_0$, closely following Cohen's (1988) approach. We begin by considering relationships among various measures of the impact of the interaction: effect size, partial regression coefficient, partial correlation, and gain in prediction (difference between squared multiple correlations or semipartial correlations). We examine sample size requirements necessary to detect the XZ interaction when X and Z are measured without error. We then explore the impact of measurement error on effect size, variance accounted for, power, and sample size requirements as a function of (a) the correlation between predictors X and Z and (b) the variance accounted for by the first order effects. Finally, the results of recent simulation studies of power of tests of interactions are presented.

Statistical Power Analysis

The power of a statistical test is the probability that the test will detect an effect in a sample when, in fact, a true effect exists in the population. As shown by Cohen (1988), the power of the statistical test depends on several parameters:

1. The specific statistical test that is chosen (e.g., parametric tests that use all available information are more powerful than nonparametric tests);
2. the level of significance chosen (e.g., $\alpha = .01$, or .05);
3. the magnitude of the true effect in the population;
4. sample size (n).

Cohen (1988) has suggested that .80 is a good standard for the minimum power necessary before undertaking an investigation. This suggestion has been accepted as a useful rule of thumb throughout the social sciences.

In considering the power for the XZ interaction term in an equation containing X and Z, we will treat the first order terms X and Z as a "set" of variables: set M for first order ("main") effects. The XZ term will constitute a second "set" I for interaction.[4] With Y as the criterion, we define the following terms:

$r^2_{Y.\mathrm{MI}}$: squared *multiple correlation* resulting from combined prediction by two sets of variables M and I, where M consists of X and Y, and I consists of XY

$r^2_{Y.\mathrm{M}}$: squared multiple correlation resulting from prediction by set M only

$r^2_{Y(\mathrm{I.M})}$: the squared semipartial (or part) correlation of set I with the criterion; $r^2_{Y(\mathrm{I.M})} = r^2_{Y.\mathrm{MI}} - r^2_{Y.\mathrm{M}}$. This is the gain in the squared multiple correlation due to the addition of set I (the interaction) to an equation containing X and Z (set M). Put otherwise, it is the proportion of total variance accounted for by set I, over and above set M.

$r^2_{Y\mathrm{I.M}}$: the squared partial correlation of set I with the criterion, or the proportion of *residual* variance after prediction by set M that is accounted for by set I,

$$r^2_{Y\mathrm{I.M}} = \frac{r^2_{Y.\mathrm{MI}} - r^2_{Y.\mathrm{M}}}{1 - r^2_{Y.\mathrm{M}}}$$

f^2: effect size for set I over and above set M, where *effect size* is defined (Cohen, 1988) as the strength of a particular effect, specifically the proportion of *systematic variance* accounted for by the effect relative to *unexplained variance* in the criterion:

$$f^2 = \frac{r^2_{Y.\mathrm{MI}} - r^2_{Y.\mathrm{M}}}{1 - r^2_{Y.\mathrm{MI}}}$$

The reader should note that the numerators of the squared partial correlation and effect size are identical and are equal to $r^2_{Y(\mathrm{I.M})}$, the squared semipartial correlation. However, the denominators of the squared partial correlation, $r^2_{Y\mathrm{I.M}}$, and the effect size, f^2, differ. The denominator of $r^2_{Y\mathrm{I.M}}$ is the residual variance after prediction from set M; the denominator of f^2 is the residual variance after prediction from sets M and I. Most importantly, the reader should note that the squared semipartial correlation, or gain in squared multiple correlation with the addition of set I to set M, is not linearly related to effect size. It is the effect size f^2 (or the closely related squared *partial* correlation, $r^2_{\mathrm{YI.M}}$) that is directly related to statistical power and not the squared semipartial correlation. This realization will clarify findings in the literature such as that of Evans (1985) in a Monte Carlo simulation "that an interaction term explaining 1% of the variance [squared semipartial correlation] is likely to be significant when the preceding first order effects have used up 80% of the variance in the dependent variable. A similar interaction is likely to be insignificant if the first order effects only absorb 10% of the dependent variable" (p. 317). With 80% of the variance explained by set M, a 1% increase in predictable variance has effect size .05; with 10% of the variance explained by set M, the 1% gain has effect size .01.[5]

Jaccard, Turrisi, and Wan (1990) have provided a useful table (their Table 3.1, p. 37, is partially reproduced here as Table 8.2) for determining sample sizes required to achieve power .80 for a test of *XZ* interaction at $\alpha = .05$. The rows of the table are $r^2_{Y.\mathrm{M}}$, the squared multiple correlations from set M, and the columns $r^2_{Y.\mathrm{MI}}$, the squared multiple correlation from both the "main effects" and the interaction. The entries are numbers of cases required under these conditions. For a constant difference between $r^2_{Y.\mathrm{MI}}$ and $r^2_{Y.\mathrm{M}}$, that is, a constant squared semipartial correlation $r^2_{Y(\mathrm{I.M})}$, the sample size requirements decrease systematically as $r^2_{Y.\mathrm{M}}$ increases, as can be seen on the diagonals of the table. Table 8.2 adds effect size estimates to Jaccard et al.'s (1990) sample size requirements. An examination of the diagonal entries shows that as $r^2_{Y.\mathrm{M}}$ increases, effect size increases, accounting for the systematic decrease in sample size requirements with increasing $r^2_{Y.\mathrm{M}}$. Table 8.3 explores the variation in effect size f^2 and the squared partial correlation $r^2_{Y\mathrm{I.M}}$ for constant squared semipartial correlation $r^2_{Y(\mathrm{I.M})}$ as $r^2_{Y.\mathrm{M}}$ increases, again showing the increase in effect size for constant $r^2_{Y(\mathrm{I.M})}$ with increasing $r^2_{Y.\mathrm{M}}$. The reader should not lose sight of the distinction between the squared semipartial correlation, $r^2_{Y(\mathrm{I.M})}$, and effect size, f^2, in reviewing literature on statistical power of the interaction.

Cohen (1988) has provided some useful guidelines for interpreting effect sizes in the social sciences. Effect sizes around $f^2 = .02$ or squared partial correlations $r^2_{Y\mathrm{I.M}} = .02$ are termed "small," around $f^2 = .15$ or $r^2_{Y\mathrm{I.M}} = .13$ are termed "moderate," and around $f^2 = .35$ or $r^2_{Y\mathrm{I.M}} = .26$ are termed "large." Cohen's reviews, as well as comprehensive meta-analytic reviews, have indicated that large effect sizes are rarely obtained in most areas of the literature in social science, education, and business. Using Cohen's power tables for power .80 at $\alpha = .05$, and assuming no measurement error in the predictors, the number of cases required to de-

Table 8.2
Effect Sizes Associated with Varying Combinations of $r^2_{Y.M}$ and $r^2_{Y.MI}$ (Adapted from Jaccard et al., 1990, Table 3.1)

		$r^2_{Y.MI}$								
		.05	.10	.15	.20	.25	.30	.35	. . .	.50
	.05		.06[a]	.12	.19	.27	.36	.46	. . .	.90
			143[b]	68	43	32	24	19	. . .	10
	.10			.06	.12	.20	.29	.38	. . .	.80
				135	65	41	29	22	. . .	10
$r^2_{Y.M}$	.15				.06	.13	.21	.31	. . .	.70
					127	60	39	27	. . .	13
	.20					.07	.14	.23	. . .	.60
						119	57	36	. . .	15
	.25						.07	.15	. . .	.50
							111	53	. . .	17
	.30							.08	. . .	.40
								103	. . .	22

NOTE: This table is adapted from Jaccard et al. (1990, Table 3.1), which provides sample size requirements for the test of the *XZ* interaction in the regression equation $\hat{Y} = b_1X + b_2Z + b_3XZ + b_0$ for power = .80 at α = .05. Effect size estimates have been added.

[a] Effect size

[b] Sample size required for power .80 at α = .05

Table 8.3
Variation in Sample Size Requirements, Effect Size (f^2), and Squared Partial Correlation ($r^2_{YI.M}$) at α = .05 for Constant Gain in Prediction or Semipartial Correlation [$r^2_{Y(I.M)}$]

$r^2_{Y.IM}$	$r^2_{Y.M}$	$r^2_{Y(I.M)}$	N[a]	f^2	$r^2_{YI.M}$
.30	.05	.25	24	.36	.26
.35	.10	.25	22	.38	.28
.40	.15	.25	21	.42	.29
.45	.20	.25	19	.45	.30
.50	.25	.25	17	.50	.33

[a] Required for power = .80, from Jaccard et al. (1990, Table 3.1, p. 37).

tect the XZ interaction (set I) are $n = 26$, 55, and 392, for large, moderate, and small effect sizes, respectively. Readers should note that the n of 55 for moderate effect size exceeds the majority of ns in Jaccard et al.'s (1990) complete Table 3.1, because the majority of effect sizes in that table exceed $f^2 = .15$, the value that defines a moderate effect size. Readers should not be misled to think that small sample sizes suffice to detect interaction effects of the strength typically found in the social sciences.

The Effects of Measurement Error on Statistical Power

In this section we examine the effects of measurement error in the predictors on several indices of effect size and statistical power. We have previously shown that the reliability of the product term increases with increases in the correlation between the components ($r_{X,Z}$). We have also pointed out that the percentage of variance accounted for by first order effects ($r^2_{Y.\mathrm{M}}$) impacts upon the effect size of the interaction. Hence, both $r^2_{Y.\mathrm{M}}$ and $r^2_{X,Y}$ are varied in our study.

The basis of our examination is Cohen's (1988) procedures for calculating the statistical power of tests in MR. This treatment of power assumes normal distributions of all predictors. We follow this tradition, assuming that X and Z are bivariate normal. However, the crossproduct term XZ will not be normally distributed: The product of two normally distributed variables does not have a normal distribution. Hence, our estimates of power, effect size, and variance accounted for are probably a bit high and our estimates of sample size requirements are probably a bit low relative to the true values (Jaccard et al., 1990). However, because the purpose of power calculations is to provide a "ballpark" estimate of the number of subjects required for an adequate study, this is not a serious limitation.

We begin by assuming no measurement error in the predictor variables. We use three effect sizes ($f^2 = .35, .15, .02$, or large, medium, and small), three levels of variance accounted for by the combined first order terms ($r^2_{Y.\mathrm{M}} = 0, .20, .50$), and two values of interpredictor correlation ($r_{X,Z} = 0, .50$). From these values we compute $r^2_{Y.\mathrm{MI}}$, the variance accounted for by the first order effects plus interactions, and $r^2_{Y(\mathrm{I.M})}$, the squared semipartial correlation or variance accounted for by the interaction over and above main effects. Under the further assumption that the predictors have identical correlations with the criterion (validities), that is, $r^2_{Y,X} = r^2_{Y,Z}$, we solve for the values of these validities that produce $r^2_{Y.M}$.

We then introduce measurement error by assuming that predictors X and Z and the criterion Y have reliability .80. We attenuate the correlations involved in the power analysis ($r^2_{Y,X}$, $r^2_{Y,Z}$, $r^2_{X,Z}$, $r^2_{Y.M}$, $r^2_{Y.IM}$, $r^2_{Y(I.M)}$) for measurement error under the assumptions of classical test theory (Lord & Novick, 1968) and Bohrnstedt and Marwell's (1978) work on the reliability of the product term. Finally, we recompute the effect size f^2 for the test of the interaction based on the attenuated correlations. From f^2 we recalculate statistical power, assuming that the researcher had used the sample sizes required for power .80 in the error-free case (n = 26, 55, 392 for large, moderate, and small effect sizes, respectively). For moderate and large effect sizes, the analysis is repeated with reliability of X, Z, and Y of .70.

Table 8.4 shows the effect of reduced reliability on effect sizes and variance acounted for by the interaction. Table 8.5 shows the effect of reduced reliability on the power of the test for the interaction assuming the ns necessary for the error-free case were utilized. It also shows the sample size required to produce power .80 for the interaction at $\alpha = .05$.

Effect Size and Measurement Error

Consider first the reduction in effect size for the interaction from an initial large effect size ($f^2 = 0.35$, Table 8.4a). The effect size f^2 is shown in row 1 across the six combinations of variance accounted for by the main effects ($r^2_{Y.M} = 0, .20, .50$) and interpredictor correlation ($r^2_{X,Z} = 0, .50$). If the reliabilities of predictors and the criterion are reduced to .80 (row 2 of large effect size section in Table 8.4), then the effect size decreases to half of its original value for $r^2_{Y.M} = 0$ and to one third of its original value for $r^2_{Y.M} = .50$. For moderate effect size ($f^2 = .15$, Table 8.4b) and small effect size ($f^2 = .02$, Table 8.4c) the proportional decreases in effect size are very similar. *The general pattern that emerges is that when reliabilities drop from 1.00 to .80, the effect size is reduced by a minimum of 50%; when reliabilities drop from 1.00 to .70, effect size is approximately 33% of its original size.*

Variance Accounted for and Measurement Error

The percentage of variance accounted for by the interaction term (i.e., $r^2_{Y(I.M)}$) over and above first order effects follows a similar pattern. Again consider large effect size ($f^2 = .35$) in Table 8.4. Row 4 of Table 8.4a for large effect size shows that when f^2 is *held constant* at .35, and the

Table 8.4
Impact of Reduced Reliability on Variance Accounted for $[r^2_{Y(\mathrm{I.M})}]$ and Effect Size (f^2) of the Interaction in the Regression Equation

$$\hat{Y} = b_1X + b_2Z + b_3XZ + b_0$$

$r^2_{Y.\mathrm{M}}$	0		.20		.50	
$r_{X,Z}$	0	.50	0	.50	0	.50
			a. Large Effect Size $f^2 = .35$ ($r^2_{Y\mathrm{I.M}} = .26$)			
Reliability			Actual effect size at $n = 26$			
1.00	.35	.35	.35	.35	.35	.35
.80	.15	.17	.14	.16	.11	.13
.70	.10	.12	.09	.11	.06	.08
Reliability			Actual $r^2_{Y(\mathrm{I.M})}$ at $n = 26$			
1.00	.26	.26	.20	.20	.13	.13
.80	.13	.15	.11	.12	.07	.07
.70	.09	.11	.07	.09	.04	.05
			b. Moderate Effect Size $f^2 = .15$ ($r^2_{Y\mathrm{I.M}} = .13$)			
Reliability			Actual effect size at $n = 55$			
1.00	.15	.15	.15	.15	.15	.15
.80	.07	.08	.07	.07	.05	.06
.70	.05	.06	.04	.05	.03	.04
Reliability			Actual $r^2_{Y(\mathrm{I.M})}$ at $n = 55$			
1.00	.13	.13	.10	.10	.07	.07
.80	.07	.07	.05	.06	.03	.04
.70	.04	.05	.04	.04	.02	.03
			c. Small Effect Size $f^2 = .02$ ($r^2_{Y\mathrm{I.M}} = .02$)			
Reliability			Actual effect size at $n = 392$			
1.00	.02	.02	.02	.02	.02	.02
.80	.01	.01	.01	.01	.01	.01
Reliability			Actual $r^2_{Y(\mathrm{I.M})}$ at $n = 392$			
1.00	.02	.02	.02	.02	.01	.01
.80	.01	.01	.01	.01	.01	.01

reliability of the predictors is perfect (1.00), the variance accounted for by the interaction is .26 when $r^2_{Y.M} = 0$, .20 when $r^2_{Y.M} = .20$, and .13 when $r^2_{Y.M} = .50$. *When reliability drops to .80, the variance accounted for the interaction decreases by 50%. When reliabilities are .70, the variance accounted for by the interaction is only 33% to 50% of that accounted for when reliabilities are 1.00.* This pattern is true for moderate and small effect sizes as well.

Statistical Power and Measurement Error

Table 8.5 addresses statistical power with reliabilities less than 1.00. The table is structured so that at reliability 1.00, the power for each effect size is .80. Note that the required sample sizes differ across effect sizes to produce constant power of .80. For the large effect size portion of Table 8.5, all *power* calculations are based on $n = 26$; for the moderate effect size portion of the table, on $n = 55$; for the small effect size portion, on $n = 392$. The pattern for loss of statistical power follows what we have already seen for effect size and variance accounted for. *Power is reduced by up to half by having reliabilities of .80 rather than 1.00 and is reduced up to two thirds when reliabilities drop to .70.*

Sample Size Requirements and Measurement Error

The sample size required to produce power of .80 at $\alpha = .05$ increases dramatically as reliability decreases (see Table 8.5). *When reliabilities drop from 1.00 to .80, the sample size required to reach power .80 at* $\alpha = .05$ *is slightly more than doubled. When reliabilities drop to .70, the sample size requirement is over three times higher than when reliabilities are 1.00.* For example, for moderate effect size, whereas $n = 55$ is required to detect an interaction when predictors are error free, sample sizes of over 200 may be required when predictor reliabilities are .70. The cost of measurement error on research implementation is enormous if adequate statistical power is to be achieved.

A Note on Variance Accounted for by Main Effects and Interpredictor Correlation

The greater the proportion of variance accounted for by the first order effects, the sharper is the decline in the effect sizes, variance accounted for, and power of the test for the interaction term as reliability decreases. Required sample sizes increase accordingly. We saw earlier in this chapter that as interpredictor correlation increases, the reliability of the product

Table 8.5
Impact of Reduced Reliability on Power of the Test for the Interaction and on Sample Size Required for Power .80 at α = .05 in the regression equation $\hat{Y} = b_1X + b_2Z + b_3XZ + b_0$

$r^2_{Y.M}$	0		.20		.50	
$r_{X,Z}$	0	.50	0	.50	0	.50
			a. Large Effect Size f^2 = .35 ($r^2_{Y1.M}$ = .26)			
Reliability			Power at n = 26			
1.00	.80	.80	.80	.80	.80	.80
.80	.45	.49	.41	.46	.34	.38
.70	.31	.37	.29	.34	.21	.26
Reliability			Required n for power = .80			
1.00	26	26	26	26	26	26
.80	54	47	59	52	75	64
.70	84	68	94	75	127	100
			b. Moderate Effect Size f^2 = .15 ($r^2_{Y1.M}$ = .13)			
Reliability			Power at n = 55			
1.00	.80	.80	.80	.80	.80	.80
.80	.48	.52	.44	.49	.37	.41
.70	.34	.40	.31	.36	.25	.29
Reliability			Required n for power = .80			
1.00	55	55	55	55	55	55
.80	109	99	122	108	153	132
.70	169	139	192	155	257	207
			c. Small Effect Size f^2 = .02 ($r^2_{Y1.M}$ = .02)			
Reliability			Power at n = 392			
1.00	.80	.80	.80	.80	.80	.80
.80	.51	.55	.47	.52	.39	.44
Reliability			Required n for power = .80			
1.00	392	392	392	392	392	392
.80	774	692	841	752	1056	909

term increases. Increases in $r_{X,Z}$ very slightly offset the loss of statistical power.

Some Corroborative Evidence: Simulation Studies

Our presentation of the effects of measurement error on power is based on Cohen's (1988) power calculations and assumptions of classical measurement theory. Evans (1985) has provided some corroborative evidence using Monte Carlo simulations. Effect sizes were found to be radically reduced by the addition of measurement error. Large effect sizes were reduced to medium effect sizes, just as is shown in Table 8.4. Evans reported that interactions accounting for 1% of the variance would be detectable if the variance accounted for by the first order effects $r^2_{Y.\mathrm{M}}$ were .80; the sample size in his simulation was $n = 760$ cases per sample, which is in the range of sample sizes reported in Table 8.5 as being necessary to detect an interaction having a small effect size accounting for 1% of the variance. As a final note, Evans reported a decrease by half in variance accounted for by the interaction when predictor reliabilities declined from 1.00 to .80; this is the magnitude of decrease reported in Table 8.4.

Two additional simulation studies provide useful models for the study of power in interactions (Dunlap & Kemery, 1988; Paunonen & Jackson, 1988). Dunlap and Kemery (1988) provide an extensive simulation study of the effects of measurement error on the power of tests for first order effects and interactions in small samples ($n = 30$). Criterion reliability was held constant at .70; predictor reliability for X and Z is varied (.20, .50, .80, or 1.00 for both predictors, as well as all combinations of mixed reliabilities of X and Z). Among the regression models tested were the following:

$$\text{Model 1:} \quad \hat{Y} = 0X + 0Z + 1XZ,$$

a pure interaction model, and

$$\text{Model 2:} \; \hat{Y} = 1X + 1Z + 1XZ,$$

in which each first order term and the interaction share equally in prediction.

For each model and each combination of predictor reliabilities, the proportion of significant interactions in 10,000 replications was reported. Table 8.6 provides a small sample of the results. In the table, "Observed Power" refers to the simulation results and "Cohen power" refers to the result in Table 8.5. Note that with predictor reliabilities of 1.00, the observed power approaches 1.00 in model 1 and exceeds .90 in model 2. When both predictors have reliability .80, observed power exceeds .90 in model 1 and is .69 in model 2. These power levels seem very high, given the demonstration in Tables 8.4 and 8.5—and they are. The high levels of statistical power are the direct result of the very large effect sizes that are considered in the simulation.

We computed the effect sizes (reported in Table 8.6) based on Cohen's (1988) procedure for six of the conditions explored by Dunlap and Kemery (the six combinations of models 1 and 2 with 1.00, .80, and .50 reliability). When the reliability was .80 or better, the effect sizes (f^2) ranged from .31 to 2.45, levels that are extraordinarily rare in practice in the social sciences! Finally, we calculated power for the interaction terms according to the approach we used to generate Table 8.5. The theoretical power calculations (given as "Cohen power" in Table 8.6) are similar to the observed power from the simulation.

Paunonen and Jackson (1988) also conducted a simulation study of the power for the interaction in model 2 with error-free predictors and sample

Table 8.6
Comparison of Observed Power in a Simulation (Dunlap & Kemery, 1988) and Power Calculations Based on Cohen (1988)

Predictor Reliability		*Model*					
		$Y = 0X + 0Z + 1XZ$			$Y = 1X + 1Z + 1XZ$		
ρ_{XX}	ρ_{ZZ}	*Observed*[a] *Power*	*Cohen*[b] *Power*	f^{2c}	*Observed*[a] *Power*	*Cohen*[b] *Power*	f^{2c}
1.00	1.00	→1.00	→1.00	2.45	.93	→1.00	.78
.80	.80	.93	.88	.39	.69	.80	.31
.50	.50	.54	.66	.22	.29	.31	.08

NOTE: Predictors X and Z are uncorrelated (one of the conditions of Dunlap and Kemery). The criterion Y has reliability .70 throughout, and $n = 30$ per sample.

[a] Power found in Dunlap and Kemery (1988) simulation under condition that $r_{X.Z} = 0$.

[b] Power calculated according to Cohen (1988) with appropriate disattenuation of correlations for unreliability.

[c] Effect sizes based on Cohen's (1988) formulation.

sizes of $n = 100$. They reported that the interaction was detected in 100% of 1000 simulated samples. Measurement error was added to the criterion, but the amount not specified. If we assume criterion reliability .70, equivalent to that in Dunlap and Kemery (1988), then we would expect 100% of the interactions to be detected, because the power approaches 1.00 for this test even with substantially smaller samples of $n = 30$. Even when the reliability of the criterion is .50, for $n = 100$ with perfectly measured bivariate normal and uncorrelated predictors, effect size is large ($f^2 = .33$), and power approaches 1.00. Once again, interactions are detected with 100% probability because they have very large effect sizes.

Paunonen and Jackson (1988) also provide simulations that match in structure the real world data of Morris et al. (1986), and hence are more realistic in terms of effect sizes. For these simulations, moderator effects were detected on average only 4.1% of the time with samples of size $n = 100$. Such a detection rate is associated with effect sizes substantially below small effect size $f^2 = .02$. Indeed, for one case considered by Morris et al. (1986), the effect size f^2 equalled .001, according to a reanalysis by Cronbach (1987).

Two recommendations come from this review of simulations of power of interactions. First, it would be very useful for our understanding of the simulations if authors would report effect size measures. This practice would permit comparison both across studies and with normative expectations for effect sizes in social science data. Second, in the absence of such reports, readers should compute the effect sizes studies in the simulation so that they are not misled by reports of very high power in the absence of measures of strength of the interaction.

Finally, we offer a caution regarding the interpretation of tests of regression models containing random measurement error in the predictors. Measurement error takes a greater toll on the power of interaction effects, relative to first order effects. Measurement error also appears to produce spurious first order effects but not spurious interactions (Dunlap & Kemery, 1988; Evans, 1985). Taken together, these two factors will lead to a greater apparent empirical support for theoretical predictions of main effects at the cost of support for theoretical predictors of interactions.

The Median Split Approach: The Cost of Dichotomization

A commonly used alternative to the multiple regression approach presented in this book is to perform median splits on each of the predictor variables and then to perform an ANOVA. This strategy loses information from each of the predictor variables, thus adding a new source of measurement error—error due to dichotomization. The effects of this procedure on simple correlations have been extensively studied in the literature on coarse categorization of variables (e.g., Bollen & Barb, 1981; Cohen, 1983).

Cohen (1983) provides data comparing the correlations, *t*-values, and power that could be expected when X and Y are both continuous versus when X is dichotomized. He shows analytically that, when two variables are sampled from a bivariate normal population, the value of the simple correlation coefficient r between a dichotomized predictor and continuous criterion is .798 of the value obtained when both variables are continuous. When the correlation between the variables is $r_{XY} = .20$; the *t*-test value for the dichotomized case is reduced to .78 of the *t*-value obtained when both variables are continuous; when $r_{XY} = .70$, t is reduced to .62 of the value obtained with the continuous predictor. For a moderate effect size ($r = .30$), the power of the test for $n = 80$ and $\alpha = .05$ is reduced from .78 to .55. Thus, for the single predictor case, dichotimization leads to a substantial drop in statistical power.

Similar analytical work has not been done for tests of interaction in multiple regression because of the difficulty introduced by the nonnormal distribution of the interaction term. As a first peek, we have conducted a small scale simulation to examine the loss of power for the test of the interaction. Using the population regression equation, $\hat{Y} = 2.00X + 0.00Z + 1.00XZ + .01$ in which X and Z are correlated .32, we took five samples of size 200 and estimated the regression equation. For each sample, we also performed median splits on X and Z and subjected the resulting data to a 2×2 ANOVA. The *t*-values for the test of the interaction obtained from the ANOVA based on median splits (t with m degrees of freedom $= F^2$ with $(1, m)$ degrees of freedom) were approximately .67 of the *t*-values obtained using the standard MR procedure. Although the exact amount of reduction in the the *t*-values will depend on several parameters (e.g., the size of the interaction effect), this example corroborates Cohen's (1983; Cohen & Cohen, 1983) admonitions concerning the cost of dichotomization on statistical power.

Principal Component Regression Is not a Cure for Power Woes

Morris et al. (1986) proposed principal component regression (PCR) as a more powerful approach than OLS multiple regression for the analysis of interaction effects (see Cronbach, 1987, for a clear conceptual presentation of the analysis). Morris et al. were particularly concerned about the effects on power of the high levels of multicollinearity that often occur between the interaction and its components in uncentered equations.[6] In an analysis of 12 real world data sets, they showed that OLS multiple regression detected an interaction in only one case, whereas parallel analyses using PCR found highly significant interactions in 10 of the 12 data sets. However, PCR, though accepted as a method for handling multicollinearity of first order predictors, is not appropriate for multiple regression models containing interactions. PCR works by eliminating some portion of the variance of each predictor. In PCR, the test of the interaction, according to Cronbach (1987), is based on a comparison of two multiple correlations. The first is that derived from prediction by those portions of the X, Z, and XZ retained in the analysis; the second is that derived from prediction by those portions of X and Z retained but from which the XZ product term has been partialled. This can be thought of as crediting the interaction with all predictable variation it shares with the first order terms. Such a procedure stands in direct contrast to the usual methods of apportioning variance among terms (those which apportion only unique variance to each effect, and those which apportion variance shared between the first order effects and the interaction to the first order effects; see Overall & Spiegel, 1969). Hence, theoretically the interaction effect should be overestimated in PCR. In a simulation, Paunonen and Jackson (1988) showed that with a nominal $\alpha = .05$ for the test of the interaction, the observed α rate was .377. With simulated data structured to match the real world data of Morris et al. (1986), significance was found in 61% of cases with PCR, but only 4% of cases with OLS. PCR does not provide an acceptable method for improving the power of tests of interactions in multiple regression (see also Dunlap & Kemery, 1987).

Coming Full Circle

In this chapter we have explored in detail the issues of reliability of measures and statistical power, as they relate to the detection of interactions in multiple regression. Measurement error severely lowers the statistical power of the test of the interaction and dramatically increases the sample size required to detect interactions. In simulation studies where evidence of high power for interactions has been reported, effect sizes have been substantially larger than those experienced in practice. For example, Chaplin (in press-a, in press-b) reviews literature on moderator effects in three broad areas of psychology with a clear finding: Observed effect sizes for interactions are very small, accounting for about 1% of variance in outcomes. Similarly, Champoux and Peters (1987) report average percentages of variance accounted for by interactions in the job design literature of 3%. To detect such effects, very large samples are required. Using large samples will ameliorate problems of power that are produced by measurement error. Large sample sizes will not, however, decrease bias in regression coefficients that is produced by measurement error. The social scientist is forewarned.

Summary

Chapter 8 considers the effects of random measurement error in the predictor variables on the bias and power of tests of the terms in the regression equation. Classical measurement theory is reviewed and is used to illustrate the effects of measurement error on the covariance matrix of the predictors, from which estimates of regression coefficients are derived. Methods of correcting the covariance matrix for random measurement error are presented, including methods recently proposed by Hiese and Fuller. Another approach proposed by Kenny and Judd based on structural equation modeling of latent variables is outlined. The concept of statistical power is introduced and the relation of several commonly used measures in regression analysis to power is considered. The power of tests of the interactions and sample size requirements for adequate power in MR are considered both for the error-free case and for the more usual situation in which the predictors are measured with less than perfect reliability. Power tables are presented that will be useful for researchers in planning their investigations. Measurement error is shown in both our theoretical analysis and in several simulation studies to lead to an appreciable decrease in the ability of MR to detect interactions.

Notes

1. For those unfamiliar with path analysis, the basis of the analysis is typically familiar ordinary least squares regression. The terms *path coefficient* and *structural coefficient* refer to standardized and unstandardized regression coefficients, respectively (Duncan, 1975).

2. Common approaches to measuring reliability make different assumptions concerning the nature of the measures (e.g., test–retest correlations require parallel form equivalence; Cronbach's alpha requires parallel form or tau equivalence). Alwin and Jackson (1980) and Kenny (1979) present discussions of reliability of measures having varying properties.

3. The constraint on the structure of a covariance matrix is that $C(X, Z) \leq s_X s_Z$. If all off-diagonal elements meet this condition, then the covariance matrix will be either positive definite or positive semidefinite. A positive definite (PD) matrix is of full rank, that is, it has all nonzero characteristic roots, a positive determinant, and may be inverted, providing the usual OLS solution for regression coefficients: $\boldsymbol{b} = \mathbf{S}_{XX}^{-1} s_{XY}$, where $\mathbf{S}_{XX}^{-1}$ is the inverse of the predictor covariance matrix. A positive semidefinite (PSD) matrix is not of full rank; there are linear dependencies among the predictors on which the covariance matrix is based (e.g., entering three predictors that must together sum to 100 points). A PSD matrix has at least one zero characteristic root, a zero determinant, and cannot be inverted; hence there is no solution for the regression coefficients.

When a covariance matrix is adjusted for error a third condition may arise: matrix indefiniteness (Feucht, 1989). The condition $C(X, Z) \leq s_X s_Z$ is not met; there is at least one negative characteristic root and a nonzero (though negative) determinant. Such a matrix may be inverted; and hence a solution for "corrected" regression coefficients derived, even though "such a moment matrix violates the structure and assumptions of the general linear model, and is inadmissable in regression analysis." (Feucht, 1989, p. 80). Sometimes, but not always, an indefinite covariance matrix of predictors will generate negative standard errors. Whenever a corrected covariance matrix is employed, its determinant should be checked before beginning the analysis.

4. We are following Cohen's (1988) development and using similar notation for effect size and other terms; thus the reader will find Cohen's highly informative and useful treatment of power quickly accessible.

5. There is a direct relationship between the reliability of the product term ($\rho_{XZ,XZ}$) and $r^2_{Y(\mathrm{I.M})}$: $r^2_{Y(\mathrm{I.M})} = \rho_{XZ,XZ}(b_3^2 \sigma_{T_X T_Z})$, where b_3^2 is taken from $\hat{Y} = b_1 X + b_2 Z + b_3 XZ + b_0$, and $\sigma_{T_X T_Z}$ is the variance of the product of true scores T_X and T_Z. Thus the percentage of attenuation in the variance accounted for in the criterion by the product term is a direct function of product term reliability (Busemeyer & Jones, 1983).

6. PCR regression was developed by Mansfield, Webster, and Gunst (1977) to adjust for multicollinearity in predictor matrices. To begin, the characteristic roots λ_i $(i = 1, p)$ and characteristic vectors $\boldsymbol{a}_i$ $(i = 1, p)$ of the $p \times p$ covariance matrix of the predictors are determined. Component scores are then formed on the set of p principal components, by postmultiplication of the raw $(n \times p)$ data matrix $\mathbf{X}$ by the matrix of characteristic vectors, that is, $\boldsymbol{u}_i = \boldsymbol{X}\boldsymbol{a}_i$, where $\boldsymbol{u}_i$ is the vector of component scores on the ith principal component. Each $\boldsymbol{u}_i$ is a linear combination of all the predictors. Those components $(\boldsymbol{u}_1, \boldsymbol{u}, \cdots \boldsymbol{u}_k)$ $(k \leq p)$ associated with the large characteristics roots $(\lambda_1, \lambda_2, \cdots \lambda_k)$ are retained for analysis. Components associated with very small characteristic roots are deleted. The criterion Y is regressed on the k orthogonal principal components that have been retained: $\hat{Y} = d_1 u_1 + d_2 u_2 \cdots d_k u_k + d_0$. The regression coefficients from that analysis $(d_1, d_2, \cdots d_k)$ are converted into regression coefficients for the original predictors by the expression $b_{\mathrm{pcr}_i} = \boldsymbol{d}' \boldsymbol{v}_i$ where b_{pcr_i} is the principal component regression coefficient for predictor i of the original predictor set. This yields the principal components regression equation in terms of the original predictors: $\hat{Y}_{\mathrm{pcr}} = b_{\mathrm{pcr}_1} X + b_{\mathrm{pcr}_2} Z + b_{\mathrm{pcr}_3} XZ + b_{\mathrm{pcr}_0}$. The b_{pcr_i} are biased estimates but they are efficient. See Mansfield et al. (1977) or Morris et al. (1986) for the complete derivation of the analysis.

The reader may recall from Chapter 4 that centering the predictors will remove most of the correlation between the crossproduct term and its component first order predictors. Hence, for simple regression equations containing an interaction, for example, $\hat{Y} = b_1 X + b_2 Z + b_3 XZ + b_0$, multicollinearity is not the source of the low level of statistical power.

9 Conclusion: Some Contrasts Between ANOVA and MR in Practice

In this book we have shown how many of the impediments to the understanding of interactions among continuous variables can be overcome. The interpretation of first order effects and interactions within the MR framework was presented in some depth for the simple case of two variables having only linear first order effects and a linear by linear interaction. This interpretation was then extended to more complex cases having more than two interacting variables, curvilinear effects, or combinations of categorical and continuous predictor variables.

Post hoc methods for probing significant interactions by testing simple slopes, determining the crossing point of the regression lines, and graphically displaying the interaction were presented for both simple and complex regression equations. The lack of invariance of regression coefficients under linear transformation was shown to have no impact whatever upon the form or interpretation of the interaction: Simple slopes and the status of the interaction as ordinal versus disordinal with regard to a predictor remain invariant under such transformations. The gain in interpretability that results from centering the variables prior to analysis was presented. A variety of tests for exploring regression equations containing higher order effects were explained, including both global tests of hypotheses and term by term step-down procedures that permit scale-free testing of the effects in the regression equation. Finally, the effects of unreliability of the predictor variables on the bias, efficiency, and power of tests of first order and interaction terms in MR were considered in depth, and several methods of correcting for unreliability were reviewed. Taken together, the procedures developed in this book allow researchers to overcome each of the impediments that have been identified to the use of the MR approach to the testing of interactions between continuous variables.

Embedded within this book, however, are two important differences in philosophy and procedure from the standard ANOVA approach, of which the ANOVA user should be aware. First is the approach to a priori specification of a model of systematic variation. Second is the emphasis on the examination of the error structure of the data to assess the adequacy of the model. These differences reflect traditions arising from the disparate origins and continued typical areas of utilization of MR versus ANOVA. ANOVA was originally developed for the analysis of planned experiments, whereas MR was originally developed for the analysis of nonexperimental observational and survey data. Although ANOVA may be mathematically considered to be a special case of MR (Cohen, 1968; Tatsuoka, 1975), the traditions associated with the two approaches have led to important differences in the practice of researchers.

In multiple regression the researcher must necessarily specify each of the terms to be included in the regression equation. This necessary specification emphasizes the importance of letting previous theory and research guide the development of the model to be tested, a requirement that will be unfamiliar to many ANOVA researchers. It also encourages the careful examination of the existing literature to determine if other, alternative models can also be developed, which can then often be directly compared with the model preferred by the investigator.

In MR when theory is unclear about the nature of a "main" (first order) or interaction effect of some of the variables, additional terms may be introduced into the equation to represent such potential influences. Such additional terms do not bias the results if, in fact, they have no influence on the criterion in the population, and are relatively uncorrelated with other predictors, although extraneous terms do decrease the efficiency of the statistical tests of the other terms and introduce Type I error. The step-down procedures described in Chapter 6 handle the efficiency problem by testing terms when all nonsignificant higher order terms have been removed. With Cohen and Cohen (1983), we encourage researchers to be cautious in introducing theoretically unexpected terms into the equation to avoid lowering power, increasing Type I error, and increasing the complexity of understanding the results, particularly when the predictors to be introduced are highly correlated with those of theoretical interest.

In ANOVA applied to *randomized experiments*, researchers typically have not considered the precise functional form of systematic variation in model specification. The standard ANOVA analysis employs a fully saturated model in which all terms through the highest order possible are always included, whether or not these higher order effects are theoretically expected to occur. Unanticipated higher order effects are detected both during omnibus effect testing and in post hoc probing, for example, a significant interaction not expected from theory but uncovered during effect testing, or an unexpected curvilinear relationship uncovered with trend analysis where only a linear relationship was expected. Less frequently recognized by ANOVA researchers is that the failure to specify a functional form does extract a penalty in terms of efficiency, as the omnibus tests of the significance of main effects and interactions in ANOVA aggregate several different functional forms, only some of which may be of theoretical interest.

The apparent discrepancy between the need to consider model specification in ANOVA versus MR is further undermined when design considerations are introduced. ANOVA tests the aggregation of all possible functional forms of the main effects and interactions *within the constraints imposed by the sampling of the levels of each factor*. The choice of too few levels of a quantitative factor in ANOVA is the same specification error as failing to include nonlinear terms in a regression model. The choice of a 2×2 factorial design means that only linear main effects in X and Z and the linear X by linear Z interactions can be detected, just as in the simple regression equation containing an interaction we considered in Chapter 2. However, in experimental designs, misspecification can only be addressed by the redesign of the experiment and the collection of new data. Thus researchers conducting randomized factorial experiments have implicitly addressed the issue of functional form at the design phase of the research; at the analysis phase, ANOVA aggregates all possible functional forms within the constraints imposed by the design.

However, when ANOVA is employed in the situation addressed in this book in which the factors are comprised of two or more *measured* variables, the functional form problem now arises in the analysis phase of the research. The researcher must decide into exactly how many levels and where each variable must be split to represent adequately the expected functional form. The use of too few levels in ANOVA with measured variables is equivalent to the omission of a higher order term in MR and typically results in biased estimates of effects. Beyond this problem of adequate modeling of the functional form of the relationship of variables to the criterion is the problem of the omission from the analysis of an important variable or interaction that is correlated with both other predictors (factors) and the criterion (dependent variable); such an omission leads to precisely the same specification errors in ANOVA and MR applied to continuous variables (see Kmenta, 1986). In sum, specification error is no less a problem in ANOVA than in MR; rather researchers using ANOVA have dwelled less upon the problem than have MR users because the specification of the model is addressed at the design rather than the analysis phase in randomized experiments.

The MR literature has placed heavy emphasis on the assumptions underlying the analysis, particularly normality, homoscedasticity, reliability of measurement of predictors, and independence of observations. Violations of assumptions often result from misspecification of the regression model. A substantial technology has emerged for the detection of such problems within regression analysis through the examination of residuals (e.g., Atkinson, 1985; Bollen & Jackman, 1990; Daniel & Wood, 1980). These authors have offered guidelines in the use of this technology to aid in the respecification of regression models so that they more adequately fit the data. Other researchers have developed methods of identifying appropriate transformations of problematic data (e.g., temporal data with autocorrelated residuals) so that appropriate tests of the hypothesized model may be performed (Judge, Hill, Griffiths, Lutkepul, & Lee, 1982; Kmenta, 1986; McCleary & Hay, 1980). Finally, the impact of influential data points on individual regression estimates has received careful attention (e.g., Atkinson, 1985; Belsley, Kuh, & Welsh, 1980; Cook & Weisberg, 1980; Stevens, 1984; Velleman & Welsh, 1981). Alternative estimation techniques that are less subject to the vagaries of individual stray data points have been proposed (Berk, 1990; Huynh, 1982).

Such careful examination of the tenability of the model has historically rarely been practiced by ANOVA researchers because of the robustness of the procedure in terms of Type I errors to violations of assumptions in between-subject, randomized experiments with equal cell ns. However, even in this optimal case, violation of the normality and equal variance assumptions can lead to decreased efficiency in tests of the effects (Levine & Dunlap, 1982). And, if the researcher uses other than randomized, between-subject designs, violations of the assumptions can easily lead to misestimation of treatment effects (Kenny & Judd, 1986; O'Brien &

Kaiser, 1985). When researchers turn to treating *measured* variables such as those discussed herein with ANOVA, the muse of robustness cannot be invoked with the same impunity.

In sum, it has been the case that ANOVA users have spent less effort on model specification and examination of the error structure of data than have MR users. This might lead ANOVA users to steer away from MR, even when it is more appropriate, because it appears that MR requires more effort. The same care in model specification and in examination of error structures is necessary in ANOVA applied to continuous variables in order that accurate effect-size estimates be achieved. That the MR literature has dwelled so extensively on the problems of parameter estimation has resulted in the availability of statistical packages that support regression diagnostics (e.g., SAS, SPSS-X), thereby facilitating the efforts of the MR user to assure model specification accuracy and freedom from error structure violations.

We have provided a complete armamentarium with which to understand interactions among continuous variables and between categorical and continuous variables in regression analysis. We hope that with these interpretational tools at their disposal, researchers familiar with the basic techniques of MR will begin testing theoretically interesting interaction terms in their equations. We also hope that researchers originally trained in ANOVA will generalize well-learned strategies and utilize the more powerful and more appropriate multiple regression framework to test interactions between continuous variables.

Appendix A: Mathematical Underpinnings

We are all familiar with the fact that if one makes simple additive transformations on a variable (i.e., adding a constant), the variance of the variable, and its covariances and correlations with other variables, remain unchanged. Only the mean changes, by a factor of the additive constant. Thus we expect that additive transformations of predictor variables will have no effect on the outcomes of multiple regression analysis. If predictor X is replaced with a variable $X + c$ where c is a constant, all regression coefficient estimates, and the variances and covariances of these estimates, are expected to remain constant. This conclusion is true so long as the regression equation contains only first order terms. In this case, only the regression constant will be affected by changes in predictor variables.

The same pattern of invariance does not hold for product terms. If a constant is added to a variable involved in a product term, the variance of the product term as well as the covariances and correlations of that product term with other terms are changed. Thus regression analyses containing product terms are *scale dependent*. The estimates of the regression coefficients, their variances and covariances, and their standard errors are altered by changes in scale. Only the raw regression coefficient for the highest order term and its standard error remain unchanged under additive transformations (Cohen, 1978).

Bohrnstedt and Goldberger (1969) provide a straightforward demonstration of the algebraic basis of this failure of invariance. Here we show (a) how the expected value (or mean) of a crossproduct term XZ depends on the expected values (or means) of the variables, X and Z, of which it is comprised; (b) how the variance of the crossproduct XZ term depends upon the expected values of X and Z; and (c) how the covariance of a crossproduct term XZ with another variable Y depends upon the expected values of X and Z. Having shown (c), it is easy to show how the covariance of XZ with X or Z depends upon expected values of X and Z as well.

Our demonstration will follow the development of Bohrnstedt and Goldberger (1969) and work in terms of expected values. (Readers wishing a review of the algebra of expectations should see Hays, 1988, Appendix B.) Those readers who are only interested in the result of the demonstration should examine expressions A.4, A.8, and A.13 for the expected value (or mean) of a crossproduct term XZ, its variance, and its covariance with a criterion variable Y, respectively. Note that in each case, the expression contains the terms $\mathrm{E}(X)$ and $\mathrm{E}(Z)$, the expected values or means of the variables forming the crossproduct and hence the source of the scale dependence.

The Expected Value (Mean) of a Product Term

We begin with two variables X and Z, making no assumption about their distribution. Think of these as two predictors in a regression analysis. We define deviation (centered) scores for each of the predictors,

$$x = X - \mathrm{E}(X) \quad \text{and} \quad z = Z - \mathrm{E}(Z)$$

or equivalently,

$$X = \mathrm{E}(X) + x \quad \text{and} \quad Z = \mathrm{E}(Z) + z$$

Their expected values (means) are $\mathrm{E}(X)$ and $\mathrm{E}(Z)$, with variances $\mathrm{V}(X) = \mathrm{E}(x^2)$ and $\mathrm{V}(Z) = \mathrm{E}(z^2)$, and covariance $\mathrm{C}(X, Z) = \mathrm{E}(xz)$.

First, we will form the crossproduct term XZ just as would be done in a regression analysis involving interaction. We form the crossproduct of the raw scores:

$$XZ = [x + \mathrm{E}(X)][z + \mathrm{E}(Z)] \tag{A.1}$$

$$XZ = [xz + z\mathrm{E}(X) + x\mathrm{E}(Z) + \mathrm{E}(X)\mathrm{E}(Z)] \tag{A.2}$$

The expected value or mean of the crossproduct of the raw scores is then

$$\mathrm{E}(XZ) = \mathrm{E}(xz) + \mathrm{E}(z)\mathrm{E}(X) + \mathrm{E}(x)\mathrm{E}(Z) + \mathrm{E}(X)\mathrm{E}(Z) \tag{A.3}$$

But for deviation scores $\mathrm{E}(x) = \mathrm{E}(z) = 0$, so the expected value of the crossproduct is as follows:

$$\mathrm{E}(XZ) = \mathrm{C}(X, Z) + \mathrm{E}(X)\mathrm{E}(Z) \tag{A.4}$$

which is the same as expression (3) of Bohrnstedt and Goldberger (1969). As we would expect, the *mean of the crossproduct depends in an orderly way on the means of the two variables*. This expression holds regardless of the distributions of X and Z.

Variance of a Product Term

To find the variance of the crossproduct term, we evaluate the expression

$$\mathrm{V}(XZ) = [XZ - \mathrm{E}(XZ)]^2 \tag{A.5}$$

substituting expression A.2 for XZ and A.4 for $\mathrm{E}(XZ)$. First, we form the square as follows:

$$\begin{aligned} \mathrm{V}(XZ) &= \{XZ - [\mathrm{C}(X, Z) + \mathrm{E}(X)\mathrm{E}(Z)]\}^2 \\ &= [xz + x\mathrm{E}(Z) + z\mathrm{E}(X) \\ &\quad + \mathrm{E}(X)\mathrm{E}(Z) - \mathrm{C}(X, Z) - \mathrm{E}(X)\mathrm{E}(Z)]^2 \end{aligned} \tag{A.6}$$

Now we take expectations, obtaining the following expression for $\mathrm{V}(XZ)$:

$$\begin{aligned} \mathrm{V}(XZ) &= \mathrm{V}(Z)\mathrm{E}^2(X) + \mathrm{V}(X)\mathrm{E}^2(Z) + \mathrm{E}(x^2z^2) + 2\mathrm{E}(X)\mathrm{E}(xz^2) \\ &\quad + 2\mathrm{E}(Z)\mathrm{E}(x^2z) + 2\mathrm{C}(X, Z)\mathrm{E}(X)\mathrm{E}(Z) - \mathrm{C}^2(X, Z) \end{aligned} \tag{A.7}$$

This expression can be simplified if we assume that X and Z are bivariate normal; this is the usual assumption in regression analysis. If variables X, Z, and W are multivariate normal, then all odd moments (first, third, fifth, etc.) are zero, (e.g., $\mathrm{E}(x) = \mathrm{E}(xzw) = \mathrm{E}(x^2z) = \mathrm{E}(x^2z^2w) = 0$). Moreover, $\mathrm{E}(x^2z^2) = \mathrm{V}(X)\mathrm{V}(Z) + 2\mathrm{C}^2(X, Z)$. Then equation A.7 simplifies to

$$\begin{aligned} \mathrm{V}(XZ) &= \mathrm{V}(Z)\mathrm{E}^2(X) + \mathrm{V}(X)\mathrm{E}^2(Z) + 2\mathrm{C}(X, Z)\mathrm{E}(X)\mathrm{E}(Z) \\ &\quad + \mathrm{V}(X)\mathrm{V}(Z) + \mathrm{C}^2(X, Z) \end{aligned} \tag{A.8}$$

This is the same as expression (6) in Bohrnstedt and Goldberger. What is important to note in expression A.8 is that *$V(XZ)$ depends upon the expected values (or means) of* X and Z. If constants are added to X, Z, or both, then $\mathrm{V}(XZ)$ will change.

Covariance of a Product Term with Another Term

Consider the covariance between a crossproduct term XZ and the criterion Y in a regression analysis.

$$\mathrm{C}(XZ, Y) = \mathrm{E}\{[XZ - \mathrm{E}(XZ)][Y - \mathrm{E}(Y)]\} \tag{A.9}$$

where

$$Y - E(Y) = y \tag{A.10}$$

and

$$XZ - E(XZ) = xz + xE(Z) + zE(X) - C(X, Z) \tag{A.11}$$

Note that expression A.11 above is formed by taking the difference between expressions A.2 and A.4.

We multiply expressions A.10 and A.11, which yields

$$C(XZ, Y) = E[xyz + xyE(Z) + zyE(X) - yC(X, Z)]$$

We take expectations, making note that $E(xy) = C(X, Y)$, $E(zy) = C(Z, Y)$, and $E(y) = 0$, so that

$$C(XZ, Y) = E(xyz) + C(X, Y)E(Z) + C(Z, Y)E(X) \tag{A.12}$$

If we assume multivariate normality, then the third moment $E(xyz)$ vanishes, yielding the expression

$$C(XZ, Y) = C(X, Y)E(Z) + C(Z, Y)E(X) \tag{A.13}$$

Expression A.13 shows that the *covariance between a product term XZ and another variable Y depends upon the expected values of the variables involved in the product term but not of the other variable*. Translating this into regression with product variables, transforming the criterion Y by additive constants will have no effect on the regression analysis.

The Covariance of a Crossproduct with a Component

In a regression analysis containing as predictors X, Z, and XZ, we are also concerned with the covariance of the crossproduct XZ with each of its component variables. With no distributional assumptions,

$$C(XZ, X) = E(x^2z) + V(X)E(Z) + C(Z, X)E(X) \tag{A.14}$$

With X and Z bivariate normal, the expression reduces to

$$C(XZ, X) = V(X)E(Z) + C(Z, X)E(X) \tag{A.15}$$

The covariance of a crossproduct with one of its components depends upon the expected value of the two variables entering the crossproduct term.

Centered Variables

In Chapter 3 we introduced regression equations including interaction terms. We began by centering the variables, that is, setting their means to zero. Here we examine the expressions for expected values, variances, and covariances given in expression A.4, A.8, A.13, and A.15 for $E(X) = E(Z) = 0$. First, with no distributional assumptions, we substitute $E(X) = 0$ and $E(Z) = 0$ into equation A.4, obtaining

$$E(XZ) = C(X, Z), \tag{A.16}$$

that is, the mean of the crossproduct terms will equal the covariance between X and Z. Note that even if X and Z are centered, the crossproduct XZ will not usually be centered.

Second, with no distributional assumptions, we substitute $E(X) = 0$ and $E(Z) = 0$ into equation A.7, finding

$$V(XZ) = E(x^2z^2) - C^2(X, Z) \tag{A.17}$$

If we assume bivariate normality, this expression simplifies to

$$V(XZ) = V(X)V(Z) + C^2(X, Z)$$

Third, with no distributional assumptions, we substitute $E(X) = 0$ and $E(Z) = 0$ into equation A.12, obtaining

$$C(XZ, Y) = E(xyz) \tag{A.18}$$

This expression reduces under the assumption of multivariate normality to

$$C(XZ, Y) = 0 \tag{A.19}$$

This result seems surprising. It says that when *two predictors X and Z and a criterion Y are multivariate normal, the covariance between the product XZ and Y will be zero*. Does this mean that there is necessarily no interaction if X, Z, and Y are multivariate normal? Yes. Turning the logic around, if there exists an interaction between X and Z in the prediction of Y, then, necessarily, the joint distribution of X, Z, and Y is *not* multivariate normal. Recall, however, that in fixed effects multiple regression, the distributional requirement applies only to the criterion. Otherwise stated, only the measurement error in the criterion must be normally distributed. Thus the result in equation A.19 does not present a problem for significance testing.

In regression analysis we are very concerned with multicollinearity or very high correlation among predictors. The covariance between a crossproduct term and a component variable is substantially reduced by centering variables. With no distributional assumption, from expression A.14,

$$C(XZ, X) = E(x^2 z) \tag{A.20}$$

Assuming bivariate normality of the prediction,

$$C(XZ, X) = 0 \tag{A.21}$$

If one compares expression A.14 with expression A.20 it is seen that *the covariance between a product term and its component is dramatically reduced by centering the predictor variables*. As was shown in Chapter 4, centering versus not centering has no effect on the highest order interaction term in multiple regression with product variables. However, centering may be useful in avoiding computational difficulties.

Appendix B: Algorithm for Identifying Scale-Independent Terms

Using an hierarchical step-down procedure to simplify regression equations with higher order terms, as was recommended in Chapter 6, requires that at each step the scale-independent term(s) be identified. The algebraic strategy presented in Chapter 3 to examine the effect of additive transformations of the variables can be used to identify the scale-independent terms in any regression equation. We present here a five-step algorithm that identifies the scale-free terms to be tested. The algorithm is illustrated using equation 5.4:

$$\hat{Y} = b_1 X + b_2 X^2 + b_3 Z + b_4 XZ + b_5 X^2 Z + b_0 \tag{5.4}$$

Although additive transformations are examined here, the procedure can be generalized to multiplicative transformations such as those involved in standardization (see Cohen, 1978).

Step 1. State the full regression equation under consideration, here equation 5.4.

Step 2. Rewrite the equation in terms of transformed predictor variables. For additive transformations, use $X' = X + c$ and $Z' = Z + f$, or equivalently $X = X' - c$ and $Z = Z' - f$. Substituting the expressions for the transformations into equation 5.4 produces the following result:

$$\begin{aligned}\hat{Y} = {} & b_1(X' - c) + b_2(X' - c)^2 + b_3(Z' - f) \\ & + b_4(X' - c)(Z' - f) + b_5(X' - c)^2(Z' - f) + b_0 \end{aligned} \tag{B.1}$$

Step 3. Expand and collect the terms of the regression equation with transformed variables:

$$\hat{Y} = (b_1 - 2b_2c - b_4f + 2b_5cf)X' + (b_2 - b_5f)X'^2 + (b_3 - b_4c + b_5c^2)Z' + (b_4 - 2b_5c)X'Z' + b_5X'^2Z' + (b_0 - b_1c + b_2c^2 - b_3f + b_4cf - b_5c^2f) \quad \text{(B.2)}$$

or equivalently

$$\hat{Y} = b_1'X' + b_2'X'^2 + b_3'Z' + b_4'X'Z' + b_5'X'^2Z' + b_0' \quad \text{(B.3)}$$

Step 4. Enumerate the relationship between each original coefficient and its corresponding transformed coefficient. Each coefficient b_i' in equation B.3 that would result from analysis of data transformed by the expressions $X' = X + c$ and $Z' = Z + f$ is shown in equation B.2 as a function of the coefficients of the original untransformed data ($b_i' s$) and the scaling constants (c and f). For example,

$$b_1' = b_1 - 2b_2c - b_4f + 2b_5cf \quad \text{(B.4)}$$

Appendix Table B.1 presents the full enumeration of relationships between coefficients of the original and transformed regression equations. In the portion of Table B.1 labeled "Coefficient Relationships" each coefficient of the transformed equation is shown as a function of the coefficient of the original equation plus the modifications due to transformation. The presence of terms in the columns b_1 through b_5 labeled "Modifications due to Transformation" indicates that the coefficient is scale dependent. Only the row for the b_5' coefficient contains no terms under "Modifications due to Transformation," hence only the b_5' coefficient is scale free.

Step 5. If a scale-free term is tested and found to be nonsignificant, it is dropped from the equation. For any equation resulting from deletion of higher order terms, delete the corresponding columns of "Modifications due to Transformation." For example, if the b_5 term were deleted from equation 5.4, the following equation would result:

$$\hat{Y} = b_1X + b_2X^2 + b_3Z + b_4XZ + b_0 \quad \text{(B.5)}$$

The b_5 column of "Modifications due to Transformation" would be deleted. After deletion, those coefficients of the transformed equation that show no entries under "Modifications due to Transformation," here b_2' and b_4', are scale free in the reduced equation. Thus in equation B.5, both b_2 and b_4 are scale invariant.

If a joint test of both the b_4 and b_5 terms were nonsignificant, leading both terms to be dropped from equation 5.4, then both the b_2' and b_3' coefficients show no entries under "Modifications due to Transformation." Hence b_2 and b_3 are scale invariant in equation $\hat{Y} = b_1X + b_2X^2 + b_3Z + b_0$; both coefficients may be tested for significance.

The present strategy is applicable to more complex equations such as equation 5.5 which includes two nonlinear effects and their interactions, or equations involving three variables, X, Z, and W, and their interactions such as equation 4.1. Appendix Table B.2 provides useful summary charts for determining the scale-free terms for these equations.

Appendix Table B.1
Coefficients of Higher Order Regression Equations after Linear Transformation of Original Variables for Equation 5.4

Original Equation:

$$\hat{Y} = b_1X + b_2X^2 + b_3Z + b_4XZ + b_5X^2Z + b_0$$

Transformed Equation:

$$\hat{Y} = b_1'X' + b_2'X'^2 + b_3'Z' + b_4'X'Z' + b_5'X'^2Z' + b_0'$$

Transformations: $X' = X + c$, $Z' = Z + f$

Coefficient Relationships:

Coefficient						
Transformed Equation	Original Equation	Modifications due to Transformation				
		b_1	b_2	b_3	b_4	b_5
b_1'	b_1		$-2b_2c$		$-b_4f$	$+2b_5cf$
b_2'	b_2					$-b_5f$
b_3'	b_3				$-b_4c$	$+b_5c^2$
b_4'	b_4					$-2b_5c$
b_5'	b_5					
b_0'	b_0	$-b_1c$	$+b_2c^2$	$-b_3f$	$+b_4cf$	$-b_5c^2f$

NOTE: Coefficient relationships indicate, for example, that coefficient b_1' of the transformed equation equals the value $(b_1 - 2b_2c - b_4f + 2b_5cf)$, where the b_i coefficients are taken from the original equation.

Appendix Table B.2
Coefficients of Higher Order Regression Equation after Linear Transformation of Original Variables for Equations 5.5 and 4.1, Where $X' = X + C$, $Z' = Z + f$, $W' = W + L$

a. *Two Factor Equation Containing a Quadratic × Quadratic Interaction*

Original equation: $Y = b_1X + b_2Z + b_3X^2 + b_4Z^2 + b_5XZ + b_6XZ^2 + b_7X^2Z + b_8X^2Z^2 + b_0$

Transformed equation: $Y = b_1'X' + b_2'Z' + b_3'X'^2 + b_4'Z'^2 + b_5'X'Z' + b_6'X'Z'^2 + b_7'X'^2Z + b_8Z'^2 + b_0'$

Coefficient Relationships:

Coefficient									
Transformed Equation	Original Equation	Modifications due to Transformation							
		b_1	b_2	b_3	b_4	b_5	b_6	b_7	b_8
b_1'	b_1			$-2b_3c$		$-b_5f^2$	$+b_6f^2$	$+2b_7cf$	$-2b_8cf$
b_2'	b_2				$-2b_4f$	$-b_5c$	$+2b_6cf$	$+b_7c^2$	$-2b_8c^2f$
b_3'	b_3							$-b_7f$	$+b_8f^2$
b_4'	b_4						$-b_6c$		$+b_8c^2$
b_5'	b_5						$-2b_6f$	$-2b_7c$	$+4b_8cf$
b_6'	b_6								$-2b_8c$
b_7'	b_7								$-2b_8f$
b_8'	b_8								
b_0'	b_0	$-b_1c$	$-b_2f$	$+b_3c^2$	$+b_4f^2$	$+b_5cf$	$-b_6cf^2$	$-b_7c^2f$	$+b_8cf$

Appendix Table B.2, continued

b. *Three Factor Equation Containing All Linear Terms*

(i) Original equation: $\hat{Y} = b_1X + b_2Z + b_3W + b_4XZ + b_5XW + b_6ZW + b_7XZW + b_0$

(ii) Transformed equation: $Y = b_1'X' + b_2'Z' + b_3'W + b_4'X'Z' + b_5'X'W' + b_6'Z'W' + b_7'X'Z'W' + b_0'$

(iii) Coefficient relationships:

Transformed Equation	Original Equation	b_1	b_2	b_3	b_4	b_5	b_6	b_7
b_1'	b_1				$-b_4f$	$-b_5h$		$+b_7fh$
b_2'	b_2				$-b_4c$		$-b_6h$	$+b_7ch$
b_3'	b_3					$-b_5c$	$-b_6f$	$+b_7cf$
b_4'	b_4							$-b_7h$
b_5'	b_5							$-b_7f$
b_6'	b_6							$-b_7c$
b_7'	b_7							
b_0'	b_0	$-b_1c$	$-b_2f$	$-b_3h$	$+b_4cf$	$+b_5ch$	$+b_6fh$	$-b_7cfh$

NOTE: Coefficient relationships indicate in the three factor equation, for example, that the coefficient b_1' of the transformed equation equals the value $(b_1 - b_4f - b_5h + b_7fh)$, where the b_i coefficients are taken from the original equation.

Appendix C: SAS Program for Test of Critical Region(s)

Written by Jenn-Yun Tein, Arizona State University

This program is applicable to cases comparing regression lines in which there are two groups and one continuous variable. It identifies critical regions where the two regression lines differ significantly using Potthoff's (1964) extension of the Johnson–Neyman procedure (see Chapter 7). Separate regression analyses within each of the groups provide the data necessary for input to this program.

Variables are entered in the order below separated by a space (free format). The values of each of the variables for the example in chapter 7 appear in lines 23 and 24 of the program. The program prints the name of the dependent variable, the limit of region 1 (XL1), and the limit of region 2 (XL2).

`DEPVAR` (short name of dependent variable)
`ALLN` = `N` (total N combining two groups)
`N1` = n_1 (number of subjects in group 1)
`N2` = n_2 (number of subjects in group 2)
`SXSQR1` = `SSX(1)` (sum of squares predicted in group 1)
`SXSQR2` = `SSX(2)` (sum of squares predicted in group 2)
`MEAN1` = `X(1)` (mean of predictor in group 1)
`MEAN2` = `X(2)` (mean of predictor in group 2)
`F` = $F_{2,N-4}$ (value of F from table)
`SSRES` = SS_{res} (sum of squares residual—add up values of SS_{res} from groups 1 and 2)
`B1` = $B_{1(1)}$ (slope for group 1)
`B01` = $B_{0(1)}$ (intercept for group 1)
`B2` = $B_{1(2)}$ (slope for group 2)
`B02` = $B_{0(2)}$ (intercept for group 2)

Program

```
00001 (local system Job Control Language [JCL])
00002 (local system JCL)
00003 (local system JCL)
00004 DATA JOHNNEYK:
00005 INPUT DEPVBL $ ALLN N1 N2 SXSQR1 SXSQR2 MEANX1 MEANX2 F
00006 SSRES B1 B01 B2 B02;
00007 MXSQR1 = MEANX1**2;
00008 MXSQR2 = MEANX2**2;
00009 SUM1 = (1/SXSQR1) + (1/SXSQR2);
00010 SUM2 = (MEANX1/SXSQR1) + (MEANX2/SXSQR2);
00011 SUM3 = (ALLN/(N1*N2)) + (MXSQR1/SXSQR1) + (MXSQR2/SXSQR2);
00012 SUMB1 = B1-B2;
00013 SUMBO = B01-B02;
00014 SUMB1SQ = SUMB1**2;
00015 SUMBOSQ = SUMBO**2;
00016 A = (((-2*F)/(ALLN-4)) * SSRES * SUM1) + SUMB1SQ;
00017 B = ((( 2*F)/(ALLN-4)) * SSRES * SUM2) + (SUMBO * SUMB1);
00018 C = (((-2*F)/(ALLN-4)) * SSRES * SUM3) + SUMBOSQ;
00019 SQRTB2AC = ((B**2) - (A*C))**.5;
00020 XL1 = (-B-SQRTB2AC)/A;
00021 XL2 = (-B+SQRTB2AC)/A;
00022 CARDS;
00023 SALARY 25 10 15 21768.4 6671180.4 2.40 2.99 3.47
00024          870923 122.9 27705.0 1872 18401.6
00025 PROC PRINT; VAR DEPVBL XL1 XL2;
00026 RUN;
00027 (local system JCL)
```

References

Allison, P. D. (1977). Testing for interaction in multiple regression. *American Journal of Sociology, 83*, 144–153.

Althauser, R. P. (1971). Multicollinearity and non-additive regression models. In H. M. Blalock (Ed.), *Causal models in the social sciences*. Chicago: Aldine.

Alwin, D. F., & Jackson, D. J. (1980). Measurement models for response errors in surveys: Issues and applications. In K. F. Schuessler (Ed.), *Sociological methodology*. San Francisco: Jossey-Bass.

Alwin, D. F., & Jackson, D. J. (1981). Applications of simultaneous factor analysis to issues of factorial invariance. In D. J. Jackson & E. F. Borgotta (Eds.), *Factor analysis and measurement in sociological research* (pp. 249–279). Beverly Hills, CA: Sage.

Anderson, L. R., & Ager, J. W. (1978). Analysis of variance in small group research. *Personality and Social Psychology Bulletin, 4*, 341–345.

Appelbaum, M. I., & Cramer, E. M. (1974). Some problems in the nonorthogonal analysis of variance. *Psychological Bulletin, 81*, 335–343.

Arnold, H. J., & Evans, M. G. (1979). Testing multiplicative models does not require ratio scales. *Organizational Behavior and Human Performance, 24*, 41–59.

Arvey, R. D., Maxwell, S. E., & Abraham, L. M. (1985). Reliability artifacts in comparable worth procedures. *Journal of Applied Psychology, 70*, 695–705.

Atkinson, A. C. (1985). *Plots, transformations, and regression*. Oxford, UK: Clarendon Press.

Belsley, D. A., Kuh, E., & Welsh, R. E. (1980). *Regression diagnostics: Identifying influential data and sources of collinearity*. New York: John Wiley.

Bentler, P. M. (1980). Multivariate analyses with latent variables: Causal modeling. In M. R. Rosenzweig & L. W. Porter (Eds.), *Annual Review of Psychology, 31*. Palo Alto, CA: Annual Reviews.

Bentler, P. M. (1989). *EOS: Structural equations program manual*. Los Angeles: BMDP Statistical Software.

Bentler, P. M., & Chou, C. P. (1988). Practical issues in structural modeling. In J. S. Long (Ed.), *Common problems/proper solutions: Avoiding error in quantitative research* (pp. 161–192). Newbury Park, CA: Sage.

Berk, R. A. (1990). A primer on robust regression. In J. Fox & J. S. Long (Eds.), *Modern methods of data analysis* (pp. 292–324). Newbury Park, CA: Sage.

Blalock, H. M., Jr. (1965). Theory building and the concept of interaction. *American Sociological Review, 30*, 374–381.

Bohrnstedt, G. W. (1983). Measurement. In P. H. Rossi, J. D. Wright, & A. B. Anderson (Eds.), *Handbook of Survey Research* (pp. 69–121). New York: Academic Press.

Bohrnstedt, G. W., & Carter, T. M. (1971). Robustness in regression analysis. In H. L. Costner (Ed.), *Sociological Methodology* (pp. 118–146). San Francisco: Jossey-Bass.

Bohrnstedt, G. W., & Goldberger, A. S. (1969). On the exact covariance of products of random variables. *Journal of the American Statistical Association, 64*, 325–328.

Bohrnstedt, G. W., & Marwell, G. (1978). The reliability of products of two random variables. In K. F. Schuessler (Ed.), *Sociological methodology*. San Francisco: Jossey-Bass.

Bollen, K. A. (1989). *Structural equations with latent variables*. New York: John Wiley.

Bollen, K. A., & Barb, K. H. (1981). Pearson's r and coarsely categorized measures. *American Sociological Review, 46*, 232–239.

Bollen, K. A. & Jackman, R. W. (1990). Regression diagnostics: An expository treatment of outliers and influential cases. In J. Fox and J. S. Long (Eds.), *Modern methods of data analysis* (pp. 257–291). Newbury Park, CA: Sage.

Borich, G. D. (1971). Interactions among group regressions: Testing homogeneity of group regressions and plotting regions of significance. *Educational and Psychological Measurement, 31*, 251–253.

Borich, G. D., & Wunderlich, K. W. (1973). Johnson–Neyman revisited: Determining interactions among group regressions and plotting regions of significance in the case of two groups, two predictors, and one criterion. *Educational and Psychological Measurement, 33*, 155–159.

Box, G. E. P., & Cox, D. R. (1964). An analysis of transformations (with discussion). *Journal of the Royal Statistical Society (Section B), 26*, 211–246.

Browne, M. W. (1984). Asymptotic distribution free methods in analysis of covariance structures. *British Journal of Mathematical and Statistical Psychology, 37*, 62–83.

Busemeyer, J. R., & Jones, L. E. (1983). Analyses of multiplicative combination rules when the causal variables are measured with error. *Psychological Bulletin, 93*, 549–562.

Byrne, B. M., Shavelson, R. J., & Muthén, B. (1989). Testing for the equivalence of factor covariance and mean structures: The issue of partial measurement invariance. *Psychological Bulletin, 105*, 456–466.

Campbell, D. T., & Fiske, D. W. (1959). Convergent and discriminant validation by the multitrait-multimethod matrix. *Psychological Bulletin, 56*, 81–105.

Champoux, J. E., & Peters, W. S. (1987). Form, effect size, and power in moderated regression analysis. *Journal of Occupational Psychology, 60*, 243–255.

Chaplin, W. F. (1991). The next generation of moderator research in personality psychology. *Journal of Personality, 59*, 143–178.

Chaplin, W. F. (in press). Personality, interactive relations and applied psychology. In S. Briggs, S. R. Hogan, & W. H. Jones (Eds.), *Handbook of Personality Psychology*. Orlando, FL: Academic Press.

Cleary, P. D., & Kessler, R. C. (1982). The estimation and interpretation of modifier effects. *Journal of Health and Social Behavior, 23*, 159–169.

Cobb, S. (1976). Social support as a moderator of life stress. *Psychosomatic Medicine, 38*, 300–314.

Cohen, J. (1968). Multiple regression as a general data-analytic system. *Psychological Bulletin, 70*, 426–443.

Cohen, J. (1977). *Statistical power analysis for the behavioral sciences*. New York: Academic Press.

Cohen, J. (1978). Partialed products *are* interactions; partialed vectors *are* curve components. *Psychological Bulletin, 85*, 858–866.

Cohen, J. (1983). The cost of dichotomization. *Applied Psychological Measurement, 7*, 249–253.

Cohen, J. (1988). *Statistical power analysis for the behavioral sciences* (2nd ed.). Hillsdale, NJ: Lawrence Erlbaum.

Cohen, J., & Cohen, P. (1975). *Applied multiple regression/correlation analyses for the behavioral sciences* (1st ed.). Hillsdale, NJ: Lawrence Erlbaum.

Cohen, J., & Cohen, P. (1983). *Applied multiple regression/correlation analyses for the behavioral sciences* (2nd ed.). Hillsdale, NJ: Lawrence Erlbaum.

Cook, R. D., & Weisberg, S. (1980). Characterization of an empirical influence function for detecting influential cases in regression. *Technometrics, 22*(4), 495–508.

Cramer, E. M., & Appelbaum, M. I. (1980). Nonorthogonal analysis of variance—once again. *Psychological Bulletin, 87*, 51–57.

Cronbach, L. J. (1987). Statistical tests for moderator variables: Flaws in analyses recently proposed. *Psychological Bulletin, 102*, 414–417.

Cronbach, L. J., & Snow, R. E. (1977). *Aptitudes and instructional methods*. New York: Irvington.

Daniel, C. & Wood, F. S. (1980). *Fitting equations to data* (2nd ed.). New York: John Wiley.

Darlington, R. B. (1990). *Regression and linear models*. New York: McGraw-Hill.

Domino, G. (1968). Differential predictions of academic achievement in conforming and independent settings. *Journal of Educational Psychology, 59*, 256–260.

Domino, G. (1971). Interactive effects of achievement orientation and teaching style on academic achievement. *Journal of Educational Psychology, 62*, 427–431.

Duncan, O. D. (1975). *Introduction to structural equation models*. New York: Academic Press.

Dunlap, W. P., & Kemery, E. R. (1987). Failure to detect moderator effects: Is multicollinearity the problem? *Psychological Bulletin, 102*, 418–420.

Dunlap, W. P., & Kemery, E. R. (1988). Effects of predictor intercorrelations and reliabilities on moderated multiple regression. *Organizational Behavior and Human Decision Processes, 41*, 248–258.

England, P., Farkas, G., Kilbourne, B. S., & Dou, T. (1988). Explaining occupational sex segregation and wages: Findings from a model with fixed effects. *American Sociological Review, 53*, 544–558.

Etezadi-Amoli, J., & McDonald, R. P. (1983). A second generation nonlinear factor analysis. *Psychometrika, 48*, 315–342.

Evans, M. G. (1985). A Monte Carlo study of the effects of correlated method variance in moderated multiple regression analysis. *Organizational Behavior and Human Decision Processes, 36*, 305–323.

Feucht, T. E. (1989). Estimating multiplicative regression terms in the presence of measurement error. *Sociological Methods & Research, 17*, 257–282.

Fiedler, F. E. (1967). *A theory of leadership effectiveness*. New York: McGraw-Hill.

Fiedler, F. E., Chemers, M. M., & Mahar, L. (1976). *Improving leadership effectiveness: The leader match concept*. New York: John Wiley.

Finney, J. W., Mitchell, R. E., Cronkite, R. C., & Moos, R. H. (1984). Methodological issues in estimating main and interactive effects: Examples from coping/social support and stress field. *Journal of Health and Social Behavior, 25*, 85–98.

Fisher, G. A. (1988). Problems in the use and interpretation of product variables. In J. Scott Long (Ed.). *Common problems/proper solutions: Avoiding error on quantitative research* (pp. 84–107). Newbury Park, CA: Sage.

Friedrich, R. J. (1982). In defense of multiplicative terms in multiple regression equations. *American Journal of Political Science, 26*, 797–833.

Fuller, W. A. (1980). Properties of some estimators for the errors-in- variables model. *The Annals of Statistics, 8*, 407–422.

Fuller, W. A. (1987). *Measurement error models*. New York: John Wiley.

Fuller, W. A., & Hidiroglou, M. A. (1978). Regression estimation after correcting for attenuation. *Journal of the American Statistical Association, 73*, 99–104.

Gallant, A. R. (1987). *Nonlinear statistical models*. New York: John Wiley.

Gulliksen, H. (1987). *Theory of mental tests*. Hillsdale, NJ: Lawrence Erlbaum. (Originally published by John Wiley, 1950).

Hayduk, L. A. (1987). *Structural equation modeling with LISREL: Essentials and advances*. Baltimore, MD: Johns Hopkins Press.

Hays, W. L. (1988). *Statistics* (4th ed.). New York: Holt, Rinehart, & Winston.

Heise, D. R. (1975). *Causal analysis*. New York: John Wiley.

Heise, D. R. (1986). Estimating nonlinear models. *Sociological Methods and Research, 14*, 447–472.

Herr, D. G., & Gaebelein, J. (1978). Nonorthogonal two-way analysis of variance. *Psychological Bulletin, 85*, 207–216.

Huitema, B. E. (1980). *The analysis of covariance and alternatives*. New York: John Wiley.

Huynh, H. (1982). A comparison of four approaches to robust regression. *Psychological Bulletin, 92*, 505–512.

Jaccard, J., Turrisi, R., & Wan, C. K. (1990). *Interaction effects in multiple regression*. Newbury Park, CA: Sage.

Johnson, P. O., & Fay, L. C. (1950). The Johnson–Neyman technique, its theory and application. *Psychometrika, 15*, 349–367.

Johnson, P. O., & Neyman, J. (1936). Tests of certain linear hypotheses and their applications to some educational problems. *Statistical Research Memoirs, 1*, 57–93.

Jöreskog, K. G. (1971). Simultaneous factor analysis in several populations. *Psychometrika, 36*, 409–426.

Jöreskog, K. G., & Sörbom, D. (1979). *Advances in factor analysis and structural equation modeling*. Cambridge, MA: Abt.

Jöreskog, K. G., & Sörbom, D. (1981). *LISREL 6: Analysis of linear structural relationships by the method of maximum likelihood*. Chicago: National Educational Resources.

Jöreskog, K. G., & Sörbom, D. (1988). *LISREL 7: A guide to the program and applications*. Chicago: SPSS.

Jöreskog, K. G., & Sörbom, D. (1989). *LISREL 7: User's reference guide*. Mooresville, IN: Scientific Software.

Judd, C. M., & McClelland, G. H. (1989). *Data analysis: A model comparison approach*. San Diego: Harcourt, Brace, Jovanovich.

Judge, G. G., & Bock, M. E. (1978). *The statistical implications of pre-test and Stein-rule estimates in enconometrics*. Amsterdam: North Holland.

Judge, G. G., Hill, R. C., Griffiths, W. E., Lutkepul, H., & Lee, T. C. (1982). *Introduction to the theory and practice of econometrics*. New York: John Wiley.

Kenny, D. A. (1975). A quasi-experimental approach to assessing treatment effects in the nonequivalent control group design. *Psychological Bulletin, 82*, 345-362.

Kenny, D. A. (1979). *Correlation and causality*. New York: John Wiley.

Kenny, D. A. (1985). Quantitative methods for social psychology. In G. Lindzey & E. Aronson (Eds.), *Handbook of Social Psychology* (3rd ed., Vol. 1., pp. 487-508). New York: Random House.

Kenny, D., & Judd, C. M. (1984). Estimating the nonlinear and interactive effects of latent variables. *Psychological Bulletin, 96*, 201-210.

Kenny, D. A., & Judd, C. M. (1986). Consequences of violating the independence assumption in analysis of variance. *Psychological Bulletin, 99*, 422-431.

Kirk, R. E. (1982). *Experimental design: Procedures for the behavioral sciences* (2nd ed.). Belmont, CA: Brooks/Cole.

Kmenta, J. (1986). *Elements of econometrics* (2nd Ed.). New York: Macmillan.

Lance, C. E. (1988). Residual centering, exploratory and confirmatory moderator analysis, and decomposition of effects in path models containing interactions. *Applied Psychological Measurement, 12*, 163-175.

Lane, D. L. (1981). Testing main effects of continuous variables in nonadditive models. *Multivariate Behavioral Research, 16*, 499-509.

LaRocco, J. M., House, J. S., & French, J. R. P., Jr. (1980). Social support, occupational stress, and health. *Journal of Health and Social Behavior, 21*, 202-228.

Lautenschlager, G. J. & Mendoza, J. L. (1986). A step-down hierarchical multiple regression analysis for examining hypotheses about test bias in prediction. *Journal of Applied Measurement, 10*, 133-139.

Levine, D. W., & Dunlap, W. P. (1982). Power of the F test with skewed data. Should one transform or not? *Psychological Bulletin, 92*, 272-280.

Long, J. S. (1983a). *Confirmatory factor analysis: A preface to LISREL*. Beverly Hills, CA: Sage.

Long, J. S. (1983b). *Covariance structure models: An introduction to LISREL*. Beverly Hills, CA: Sage.

Lord, F. M., & Novick, M. R. (1968). *Statistical theories of mental test scores*. Reading, MA: Addison-Wesley.

Lubin, A. (1961). The interpretation of significant interaction. *Educational and Psychological Measurement, 21*, 807-817.

Lubinski, D., & Humphreys, L. G. (1990). Assessing spurious "moderator effects": Illustrated substantively with the hypothesized ('synergistic') relation between spatial and mathematical ability. *Psychological Bulletin, 107*, 385-393.

Maddala, G. S. (1977). *Econometrics*. New York: McGraw-Hill.

Mansfield, E. R., Webster, J. T., & Gunst, R. F. (1977). An analytic variable selection technique for principal component regression. *Applied Statistics, 26*, 34-40.

Marascuilo, L. A., & Levin, J. R. (1984). *Multivariate statistics in the social sciences*. Belmont, CA: Brooks/Cole.

Marquardt, D. W. (1980). You should standardize the predictor variables in your regression models. *Journal of the American Statistical Association, 75*, 87-91.

Marsden, P. V. (1981). Conditional effects in regression models. In P. V. Mardsen (Ed.), *Linear Models in Social Research* (pp. 97–116). Beverly Hills, CA: Sage.

McCleary, R. & Hay, R. A., Jr. (1980). *Applied time series analysis*. Beverly Hills, CA: Sage.

Mooijaart, A., & Bentler, P. M. (1986). Random polynomial factor analysis. In E. Diday et al. (Eds.), *Data analysis and informatics* (pp. 241–250). Amsterdam: Elsevier Science.

Morris, J. H., Sherman, J. D., & Mansfield, E. R. (1986). Failures to detect moderating effects with ordinary least squares—moderated multiple regression: Some reasons and a remedy. *Psychological Bulletin, 99*, 282–288.

Morrison, D.F. (1976). *Multivariate statistical methods* (2nd ed.). New York: McGraw-Hill.

Mosteller, F., & Tukey, J. W. (1977). *Data analysis and regression: A second course in statistics*. Reading, MA: Addison-Wesley.

Myers, J. L. (1979). *Fundamentals of experimental design*. Boston: Allyn & Bacon.

Neter, J., Wasserman, W., & Kutner, M. H. (1989). *Applied Linear Regression Models* (2nd ed.). Homewood, IL: Irwin.

Nunnally, J. C. (1978). *Psychometric Methods* (2nd ed.). New York: McGraw-Hill.

O'Brien, R. G., & Kaiser, M. D. (1985). MANOVA method for analyzing repeated measures designs: An extensive primer. *Psychological Bulletin, 97*, 316–333.

Oldham, G. R., & Fried, Y. (1987). Employee reactions to workplace characteristics. *Journal of Applied Psychology, 72*, 75–80.

Overall, J. E., Lee, D. M., & Hornick, C. W. (1981). Comparisons of two strategies for analysis of variance in nonorthogonal designs. *Psychological Bulletin, 90*, 367–375.

Overall, J. E., & Spiegel, D. K. (1969). Concerning least squares analysis of experimental data. *Psychological Bulletin, 72*, 311–322.

Overall, J. E., Spiegel, D. K., & Cohen, J. (1975). Equivalence of orthogonal and nonorthogonal analysis of variance. *Psychological Bulletin, 82*, 182–186.

Paunonen, S. V., & Jackson, D. N. (1988). Type I error rates for moderated multiple regression analysis. *Journal of Applied Psychology, 73*, 569–573.

Pedhazur, E. J. (1982). *Multiple regression in behavioral research*. New York: Holt, Rinehart & Winston.

Peixoto, J. L. (1987). Hierarchical variable selection in polynomial regression models. *The American Statistician, 41*, 311–313.

Potthoff, R. F. (1964). On the Johnson-Neyman technique and some extensions thereof. *Psychometrika, 29*, 241–256.

Rao, C. R. (1973). *Linear statistical inference and its applications*. New York: John Wiley.

Rogosa, D. (1980). Comparing nonparallel regression lines. *Psychological Bulletin, 88*, 307–321.

Rogosa, D. (1981). On the relationship between the Johnson-Neyman region of significance and statistical tests of parallel within group regressions. *Educational and Psychological Measurement, 41*, 73–84.

Schmidt, F. L. (1973). Implications of a measurement problem for expectancy theory research. *Organizational Behavior and Human Performance, 10*, 243–251.

Simonton, D. K. (1987). Presidential inflexibility and veto behavior: Two individual-situational interactions. *Journal of Personality, 55*, 1–18.

Smith, K. W. & Sasaki, M. S. (1979). Decreasing multicollinearity: A method for models with multiplicative functions. *Sociological Methods and Research, 8*, 35–56.

Sobel, M. E. (1982). Asymptotic confidence intervals for indirect effects in structural equation models. In K. Schuessler (Ed.), *Sociological methodology.* San Francisco: Jossey-Bass.

Sockloff, A. L. (1976). The analysis of nonlinearity via linear regression with polynomial and product variables: An examination. *Review of Educational Research, 46*, 267–291.

Southwood, K. E. (1978). Substantive theory and statistical interaction: Five models. *American Journal of Sociology, 83*, 1154–1203.

Sprecht, D. A., & Warren, R. D. (1975). Comparing causal models. In D. R. Heise (Ed.), *Sociological methodology.* San Francisco: Jossey-Bass.

Stevens, J. P. (1984) Outliers and influential data points in regression analysis. *Psychological Bulletin, 95*(2), 334–344.

Stimson, J. A., Carmines, E. G., & Zeller, R. A. (1978). Interpreting polynomial regression. *Sociological Methods and Research, 6*, 515–524.

Stine, R. (1990). An introduction to bootstrap methods. In J. Fox & J. S. Long (Eds.), *Modern methods of data analysis* (pp. 325–374). Newbury Park, CA: Sage.

Stolzenberg, R.M. (1979). The measurement and decomposition of causal effects in nonlinear and nonadditive models. In K. F. Schuessler (Ed.), *Sociological methodology.* San Francisco: Jossey-Bass.

Stolzenberg, R. M., & Land, K. C. (1983). Causal modeling and survey research. In P. H. Rossi, J. D. Wright, & A. B. Anderson (Eds.), *Handbook of survey research* (pp. 613–675). New York: Academic Press.

Stone, E. F., & Hollenbeck, J. R. (1984). Some issues associated with the use of moderated regression. *Organizational Behavior and Human Performance, 34*, 195–213.

Stone, E. F., & Hollenbeck, J. R. (1989). Clarifying some controversial issues surrounding statistical procedures for detecting moderator variables: Empirical evidence and related matters. *Journal of Applied Psychology, 74*, 3–10.

Tate, R. L. (1984). Limitations of centering for interactive models. *Sociological Methods and Research, 13*, 251–271.

Tatsuoka, M. M. (1975). *The general linear model: A "new" trend in analysis of variance.* Champaign, IL: Institute for Personality and Ability Testing.

Teghtsoonian, R. (1971). On the exponents in Stevens' law and the constant in Ekman's law. *Psychological Review, 78*, 71–80.

Thomas, G. B. (1972). *Calculus and analytic geometry* (4th ed.). Reading, MA: Addison-Wesley.

Velleman, P. F., & Welsh, R. E. (1981). Efficient computing of regression diagnostics. *American Statistician, 35*, 234–242.

Wenger, B., (Ed.). (1982). *Social attitudes and psychophysical measurement.* Hillsdale, NJ: Lawrence Erlbaum.

West, S. G. & Aiken, L. S. (1990). *Conservative tests of simple effects.* Unpublished manuscript, Arizona State University, Tempe, AZ.

West, S. G., & Finch, J. F. (in press). Measurement and analysis issues in the investigation of personality structure. In S. Briggs, R. Hogan, & W. Jones (Eds.), *Handbook of Personality Psychology.* New York: Academic Press.

West, S. G., Sandler, I., Pillow, D. R., Baca, L., & Gersten, J. C. (in press). The use of structural equation modeling in generative research. *American Journal of Community Psychology.*

Winer, B. J. (1971). *Statistical principles in experimental design* (2nd ed.). New York: McGraw-Hill.

Won, E. Y. T. (1982). Incomplete corrections for regressor unreliabilities. *Sociological Methods and Research, 10*, 271–284.

Wong, S. K., & Long, J. S. (1987). *Parameterizing nonlinear constraints in models with latent variables*. Unpublished manuscript, Indiana University, Department of Sociology, Bloomington, IN.

Wonnacott, R. J. & Wonnacott, T. H. (1979). *Econometrics* (2nd ed.). New York: John Wiley.

Wright, G. C., Jr. (1976). Linear models for evaluating conditional relationships. *American Journal of Political Science. 20*, 349–373.

Yerkes, R. M., & Dodson, J. D. (1908). The relation of strength of stimulus to rapidity of habit formation. *Journal of Comparative Neurology of Psychology, 18*, 459–482

Glossary of Symbols

Symbol	Definition	Page
$\boldsymbol{a}$	characteristic vector of $\mathbf{S}_{XX}$, the variance–covariance matrix of the predictors	171
ANOReg	analysis of regression	25
ANOVA	analysis of variance	4
$b_1, b_2 \ldots b_k$	unstandardized regression coefficients based on centered predictors	1
$b'_1, b'_2 \ldots b'_k$	unstandardized regression coefficients based on uncentered predictors	30
$b^*_1, b^*_2 \ldots b^*_k$	standardized regression coefficients	43
b_0	regression constant (Y-intercept) based on centered predictors	1
b^*_0	regression constant (Y-intercept) in standardized regression equation containing an interaction	43
b_{YX}	unstandardized regression coefficient of Y on X in the single predictor case	141
$b_{YX.Z}$	unstandardized partial regression coefficient of Y on X in the two predictor (X and Z) case	142
$\boldsymbol{b}$	vector of unstandardized regression coefficients based on centered data	25
c	additive constant	30

Symbol	Definition	Page
Cov_{XZ}	covariance between X and Z	44
$C(X, Y)$	covariance between X and Y	141
CV_Z	conditional value of Z, the value of Z at which the simple regression of Y on X is considered	18
CV_W	conditional value of W, the value of W at which the simple regression of Y on X is considered	54
$\mathbf{d}$	vector of regression coefficients from principal component regression	171
df	degrees of freedom	6
D_i	dummy code for identifying group membership	117
E_i	effect code for identifying group membership	128
f	additive constant	30
f^2	effect size	157
G	number of levels of a categorical (group) variable	117
I_i	intercept for group i in slope/intercept computation	126
k	number of predictors in a regression equation, not including the regression constant b_0	16
MR	multiple regression	3
$\text{MS}_{Y-\hat{Y}}$	mean square residual from analysis of regression	25
n	number of cases in a sample	16
OLS	ordinary least squares	96
PCR	principal component regression	168
$r^2_{Y(\text{I.M})}$	squared semi-partial (or part) correlation of set I with the criterion with set M partialled out	157
$r^2_{Y\text{I.M}}$	squared partial correlation of set I with the criterion with set M partialled out	157
$r^2_{Y.\text{M}}$	squared multiple correlation resulting from prediction of criterion by a set of variables M	157
$r^2_{Y.\text{MI}}$	squared multiple correlation resulting from combined prediction of criterion by a set of variables M plus their interaction I	157
R^2_{in}	in hierarchical regression in which predictor j is added to predictor i to predict the criterion, the	106

Symbol	Definition	Page
	squared multiple correlation with both predictor i and j in the equation	
R^2_{out}	in hierarchical regression in which predictor j is added to predictor i to predict the criterion, the squared multiple correlation with only predictor i in the equation and j out of the equation	106
r_{XX}, r_{ZZ}	sample estimate of population reliability ρ_{XX} of predictor X, of population reliability ρ_{ZZ} of predictor Z, respectively	142
r_{XY}	zero order correlation between X and Y	142
$r_{X,Y}$	zero order correlation between X and Y	160
$\mathbf{S}_b$	sample variance covariance matrix of regression coefficients	25
s_b	standard error of a simple slope	16
s_d	standard error of the difference between two simple slopes	20
s_i	standard deviation of predictor i	45
S_i	slope for group i in slope/intercept computation	126
s_{ii}	sample variance of unstandardized regression coefficient b_i; ith diagonal element of $\mathbf{S}_b$, for centered predictors	16
s_{ij}	sample covariance between unstandardized regression coefficients b_i and b_j; off-diagonal element of $\mathbf{S}_b$ for centered predictors	16
s'_{ij}	sample covariance between unstandardized regression coefficients b'_i and b'_j; off-diagonal element of $\mathbf{S}_b$ for uncentered predictors	34
s_{L}, s_{M}, s_{H}	standard errors of simple slopes of Y on X at Z_{L}, Z_{M}, and Z_{H}, respectively (centered case)	17
s'_{L}, s'_{M}, s'_{H}	standard errors of simple slopes of Y on X at Z'_{L}, Z'_{M}, and Z'_{H}, respectively (uncentered case)	44
s^*_{L}, s^*_{M}, s^*_{H}	standard errors of simple slopes of Y on X at values of standardized $Z = -1.0$, 0.0, and 1.0, respectively, (i.e., one standard deviation below the mean of standardized predictor Z, at the mean of standardized predictor Z, and one standard deviation above the mean of standardized predictor Z)	46

Symbol	Definition	Page
$\mathbf{S}_{XX}$	sample variance–covariance matrix of predictors	25
$\mathbf{s}_{XY}$	vector of covariances of each predictor with the criterion	171
s_Y	standard deviation of criterion Y	45
T_X	true score on variable X	140
U	a linear combination of regression coefficients	25
$\boldsymbol{u}_i$	vector of component scores on principal component i of $\mathbf{S}_{XX}$	171
w_j	weight applied to regression coefficient j in forming a simple slope	25
W	centered predictor W (in deviation form)	49
$\boldsymbol{w}$	weight vector applied to vector of regression coefficients to form a simple slope	25
WABOVE	$(W - \mathrm{CV}_W)$, where $\mathrm{CV}_W = 1$ standard deviation of W	58
WBELOW	$(W - \mathrm{CV}_W)$, where $\mathrm{CV}_W = -1$ standard deviation of W	58
W_{CV}	$(W - \mathrm{CV}_W)$, predictor W from which a conditional value CV_W has been subtracted	54
$W_{\mathrm{L}}, W_{\mathrm{H}}$	score on centered predictor W one standard deviation below mean of W, one standard deviation above mean of W, respectively	51
X	centered predictor X (in deviation form)	1
$\overline{X}$	sample mean of X	11
X'	uncentered predictor X	30
X^*	latent variable X in structural equation model	153
X_{cross}	the value of X at which two simple regression lines of Y on X at values of Z cross, for centered X	24
X'_{cross}	the value of X at which two simple regression lines of Y on X at values of Z cross, for uncentered X	32
XZ	the crossproduct of centered X and centered Z	2
$X'Z'$	crossproduct of uncentered predictors X' and Z'	30
X^*Z^*	product of latent variables X^* and Z^* in structural equation model	153
X^2Z	crossproduct of square of centered X with Z	6

Symbol	Definition	Page
XZW	crossproduct of three centered predictors X, Z, and W	49
XZABOVE	product of X with ZABOVE	19
XZBELOW	product of X with ZBELOW	19
$\hat{Y}$	predicted score in unstandardized regression equation	1
Z	centered predictor Z (in deviation form)	1
Z'	uncentered predictor Z	30
Z^*	latent variable Z in structural equation model	153
ZABOVE	$(Z - CV_Z)$, where $CV_Z = 1$ standard deviation of Z	19
ZBELOW	$(Z - CV_Z)$, where $CV_Z = -1$ standard deviation of Z	19
Z_{cross}	the value of Z at which two simple regression lines of Y on Z at values of X cross, for centered Z	24
Z_{CV}	$(Z - CV_Z)$, predictor Z from which a conditional value CV_Z has been subtracted	18
Z_L, Z_M, Z_H	scores on centered predictor Z one standard deviation below the mean of Z, at the mean of Z, and one standard deviation above the mean of Z, respectively	13
Z'_L, Z'_M, Z'_H	score on uncentered predictor Z' one standard deviation below the mean of Z', at the mean of Z', and one standard deviation above the mean of Z', respectively	33
z_X, z_Z	standardized predictor X and Z, respectively, based on centered X and Z	43
$z_{X'}, z_{Z'}$	standardized predictor X and Z, respectively, based on uncentered X and Z	43
$z_X z_Z$	crossproduct of z-scores on centered predictors X and Z	43
$\hat{z}_Y$	predicted standardized score from standardized regression equation	43
ϵ_i	residual $(Y - \hat{Y})$	25
ϵ_X	measurement error in observed score on variable X	140

Symbol	Definition	Page
λ_i	factor loading in measurement model (structural equation modeling)	152
λ_i	characteristic roots of covariance matrix of predictors	171
μ_X, μ_Z	population means of *X* and *Z*, respectively	144
ρ_{XX}, ρ_{ZZ}	population reliabilities of *X* and *Z*, respectively	141
ρ_{XY}	population zero order correlation between *X* and *Y*	142
$\rho_{XZ,XZ}$	population reliability of crossproduct term *XZ*	144
Σ_b	population variance–covariance matrix of the regression coefficients	25
σ_b^2	population variance of a simple slope	25
σ_ϵ^2	population variance of the residuals, $\Sigma(Y - \hat{Y})^2$	25
$\sigma_{\epsilon_X}^2$, $\sigma_{\epsilon_Z}^2$	population variance of the measurement errors on predictor *X* and *Z*, respectively	141
$\sigma_{\epsilon_{XZ}}^2$	population variance of the measurement errors on crossproduct *XZ*	147
σ_{ij}	population covariance between two regression coefficients b_i and b_j	26
σ_{jj}	population variance of regression coefficient *j*	26
σ_{Tx}^2	variance of true scores on predictor *X*	141
σ_X^2, σ_Z^2	population variance of predictors *X*, *Z*, respectively	44
σ_{XZ}^2	population variance of crossproduct term *XZ*	44

Author Index

Subject Index

About the Authors

Leona S. Aiken is currently Professor of Psychology at Arizona State University. She previously held positions as Professor and Director of the Quantitative Psychology program and Senior Study Director at the Institute for Survey Research at Temple University. During 1991–1992 she will be a visiting scholar in quantitative psychology at UCLA. Her administrative posts have included Associate Provost of Temple University and Associate Dean at Arizona State University. Her primary research interests are in the statistical analysis of large scale outcome evaluations as well as health psychology, the latter focusing on the development and evaluation of programs to increase medical compliance. Commencing in 1991, she will serve as associate editor of the *Journal of Personality*.

Stephen G. West is currently Professor of Psychology and co-principal investigator of the Prevention Intervention Research Center at Arizona State University. He was previously at Florida State University and has held visiting faculty positions at University of Wisconsin, University of Texas, Duke University, University of Heidelberg, and UCLA. He is currently editor of the *Journal of Personality*, was associate editor of *Evaluation Review* and was co-editor of *Evaluation Studies Review Annual* (Vol. 4). His primary research interests are in the design and statistical analysis of field research and applied social psychology.

Raymond R. Reno received his Ph.D. from Arizona State University. He is currently Assistant Professor of Psychology at Notre Dame University. His primary research interests involve the effects of self-focused attention and the influence of norms on behavior, as well as statistical applications in the social sciences.